THE SMOKE OF THE GODS

ALSO BY ERIC BURNS

NONFICTION

Broadcast Blues

The Joy of Books

The Spirits of America: A Social History of Alcohol

Infamous Scribblers: The Founding Fathers and the Rowdy Beginnings of American Journalism

FICTION

The Autograph

ERIC BURNS

The Smoke of the Gods

A Social History of Tobacco

TEMPLE UNIVERSITY PRESS
Philadelphia

TEMPLE UNIVERSITY PRESS
1601 North Broad Street
Philadelphia PA 19122
www.temple.edu/tempress

Published 2007
Printed in the United States of America

Text design by Kate Nichols

∞

The paper used in this publication meets the requirements
of the American National Standard for Information Sciences—Permanence
of Paper for Printed Library Materials, ANSI Z39.48-1992

Library of Congress Cataloging-in-Publication Data

Burns, Eric.
The smoke of the gods : a social history of tobacco / Eric Burns.
p. cm.
Includes bibliographical references and index.
ISBN 1-59213-480-7 (hardcover : alk. paper)
1. Tobacco use—History. 2. Tobacco—Social aspects.
3. Tobacco—United States—History. 4. Smoking—History.
5. Antismoking movement—History.
6. Antismoking movement—United States—History. I. Title.

HV5730.B87 2007
394.1'4—dc22 2006017390

2 4 6 8 9 7 5 3 1

To

TIM AND VERA COLE
AND
BURGESS RUSSELL

for friendship
transcending distance

CONTENTS

INTRODUCTION • The Ancient World • 1

ONE • The Old World • 14

TWO • The Enemies of Tobacco • 36

THREE • The Politics of Tobacco • 56

FOUR • The Rise of Tobacco • 71

FIVE • Rush to Judgment • 97

SIX • Ghost, Body, and Soul • 103

SEVEN • The Cigarette • 128

EIGHT • The Carry Nation of Tobacco • 140

NINE • The Last Good Time • 168

TEN • The Case against Tobacco • 189

ELEVEN • The Turning Point • 216

EPILOGUE • "The Ten O'Clock People" • 240

Acknowledgments • 245

Notes • 247

Select Bibliography • 259

Index • 265

INTRODUCTION

The Ancient World

IMAGINE YOURSELF A MAYA. You are short, muscular, and well conditioned; your skin is dark, and your hair is long and unruly. You live in the Yucatan Peninsula of Mexico more than 1,500 years ago, surrounded by forest and warmed by a sun that is almost always set to low-bake. You raise crops and kill animals, trade pelts and build roads, play a primitive version of basketball, and dance to the rhythms of impromptu percussion. Your house is a hut with mud-covered walls and a roof of palm leaves.

You have filed your teeth until the tips are sharp. You have not only pierced your ears but stretched them far beyond nature's intent by wedging ever larger plugs into the holes in the lobes; they are now the size of small pancakes. There is a tattoo of the sun on your forearm and of a mountain peak on your shoulder; other tattoos, in other locations, are merely designs, representing your own notions of beauty and symmetry rather than the shapes of nature. Because your parents held a small piece of coal tar at the bridge of your nose when you were an infant and forced you to focus on it, your eyes are crossed in a most becoming manner, reminiscent of the sun god. You also have you parents to thank for the rakish slope of your forehead; while your mother showed you the coal tar, your father was angling a piece of wood onto the top of your face and pressing down as hard as he could for as long as he could, this so that your still plaint skull would be streamlined according to the day's fashion. Cosmetic surgery, pre-Columbian style.

On special occasions you paint yourself red. You decorate your clothes with feathers plucked from birds that have been specially bred for the extravagance of their plumage. You hang tiny obsidian mirrors from your hair, and they tinkle as you walk, catching the sun. The ring in your nose also catches the sun. So do your beetle-wing necklace and the pieces of jade in those overgrown lobes of yours. As you make your way through the village at the peak of noon, your head is a symphony of reflected light.

You are a man of substance as well as style. Your tools are crude, it is true, and although the Maya have joined other peoples in discovering the wheel, it is to you but an aid to various minor tasks, not a civilization-altering means of transportation. Nor do you farm wisely. You clear the land by burning away existing vegetation, which in time depletes the soil and makes it necessary for you to clear more land, ever moving on, ever destroying. Your cities are primitive, even by the standards of the time, and education is for the few.

But your mathematicians conceived the zero more than a thousand years before the Hindus, and your astronomers devised a calendar that is the model of precision, enabling you to understand the phases of the moon and to plot the frequency of solar eclipses, which used to seem random. Your architecture will stand the test of time, as will some of your sculpture and the glyphs that constitute your written language. Your government is not sophisticated, but neither is it repressive, and you are not easily drawn into arguments with your neighbor or warfare with other tribes in the region. Hardships are uncommon and, for the most part, manageable.

In his *Institutes of the Christian Religion*, John Calvin will one day write that "a sense of divinity is by nature engraven on human hearts." This is certainly true for you and your fellow Maya, to whom the world is a place of mystery and foreboding, despite the successes of your science. Sometimes you look up at a blue sky, other times at a gray one, and you cannot predict the change. Sometimes the air is still, sometimes it blows fiercely, and you do not know the cause of either. It may rain so hard that your hut is destroyed or so little that your corn and squash do not grow, and no one can say why.

And it is not just the weather that puzzles you. Almost daily there are sights you do not recognize and sounds you cannot identify and smells that suggest the proximity of danger. Even the simplest things can appear ominous when the explanation is hidden. One morning the hunt goes well, the next there are no animals to be seen for miles. One afternoon your body feels fine, the next you are fevered or stiff or sore. As the years pass, your reflexes slow, your vision dims; what you could once accomplish with ease now seems a struggle, and you do not understand why you are no longer the same. Death may take you without warning, as it does the elders of your tribe—but where? Is death a physical place, like a body of water to which one journeys for refreshment, or simply an end to all you know and ever will know, like a sleep from which one never awakens? The night is not just the other side of day, but in the purity and depth of its blackness, it is a perfect metaphor for your relationship to the cosmos. No wonder you take refuge in Calvin's "sense of divinity." Nature is an enigma that demands consolation.

Your priests have painted themselves blue. Their fabrics are richer than yours and more profusely feathered, especially those of the chief priest, or *halach uinic.* You do not begrudge him this. He is a man you respect greatly. He is also a man whose services you cannot do without, as he speaks to the immortals on behalf of the entire tribe. From the great Chacs he wants a rich harvest; from the esteemed Ah Puch he requests a long life for all; from the grandmother of the sun he asks only that there be a dawn each day to wash the night from the sky.

You sit before him in the temple and hold up your pipe, which a later historian will refer to as a "portable altar" and another will call the "most ingenious religious artifact ever invented." It is made of clay or stone, and the bowl is shaped like the head of a god. You fill it with the leaves from a plant known as tobacco, which you believe to be the gods' gift to you, one of the surest signs they have given both of their existence and their regard for human beings, and you light it from the fire that the priests have built near the altar. You draw on it and gulp in the smoke, bringing it deep into your body and holding it for several seconds, allowing it to suffuse your innards, seep through every byway, into every corner and over every organ. You blow out, watching intently as the smoke drifts up toward the sun and the clouds. Again you take it in, again you blow it out, and then again and again, one more time and another, as the regularity of inhaling and exhaling, holding and releasing, becomes a kind of silent incantation.

But you are not just smoking. You are praying, and the smoke that you expel is the emissary of your prayers. It is incense with a mission. Yes, it rises only a few feet before disappearing, but you are not discouraged; to the contrary, the very fact that the smoke is here one moment and gone the next strikes you as magical, evidence that the gods are reaching down to accept your pleas, gathering them into celestial realms with invisible hands. Or perhaps they are breathing in as you breathe out, a kind of give-and-take that enables you to communicate with them via second-hand smoke. This does not necessarily mean that your harvest will be bountiful or your days of great number or even that the progress of the dawns is certain. There are tribes to the north of you that blow smoke over their weapons before battle; this does not ensure accuracy. There are other tribes that blow smoke down the throats of the animals they have killed; this does not ensure protection from the creatures' ghosts.

Still, you believe that the vanished emissions mean the gods have taken your prayers under advisement, are treating them seriously. It is all a person can hope for in a capricious world.

Before long, the tobacco has its way with you. You begin to relax and, at the same time, to feel more alert, more receptive to voices both inner and otherwise. You expect something out of the ordinary to happen and prepare yourself for the best. "Plants whose properties place the user in an unusual state," writes the historian Francis Robicsek, "have always been looked upon by the natives of the New World as endowed with supernatural powers." It is for this reason that so many other tribes of your time employ alcoholic beverages and hallucinogenic flora, confusing a drunken stupor for a pietistic trance and converting the mistake into an article of faith.

But not you. Not any of the Maya. As far as you are concerned, the state induced by beer and wine and peyote is *too* unusual. A worshipper becomes so besotted that he cannot even remember the name of his deity, much less the specific request he means to make of him. The ritual defeats, rather than assists, its goal.

This does not happen with tobacco. Some of your smaller blood vessels contract when you smoke, but not much. A few of the larger ones expand, but only a bit. The temperature of your skin drops, but so little that you cannot really notice it. Your digestive processes slow, your pulse quickens; all of it, though, is minimal. The effect, in other words, is a mild one, although it is probably more narcotizing than that of today's tobacco, which loses some of its alkaloid-induced impact in the process of curing and manufacture.

What it all means is that your increased attentiveness may or may not be an illusion. It does not matter, though, for an illusion that cannot be disproved bears a striking resemblance to the truth. These prayers of yours, these monologues on the nature of existence and your particular place in it, are impossible to distinguish from a satisfying dialogue with the gods. This is what tobacco does for you; it is how the plant confuses you and why your faith depends on it.

You and your countrymen are not the first to know tobacco. Other tribes came across the plant as long as 16,000 years before Christ, and the natives of Peru and Ecuador were cultivating it between 5,000 and 3,000 B.C. But these early peoples did not ritualize its use as you do. They did not attach such meaning to it, bestow upon the leaf such an exalted role in the culture.

And so, painted and plumed and tattooed, sharp-toothed and large-lobed, proud of your material achievements yet in almost trembling awe of the world around you, you sit in your temple in a Mexican jungle in the distant past and smoke your pipe reverently, casting your eyes upward and thinking your most important thoughts. You do not know that you are poisoning yourself. You do not know that the gods who gave you tobacco are the gods of the underworld. What you do know is that if your prayers are not answered today, there is tobacco enough for all the tomorrows you can imagine.

YOU BELIEVE THAT disease is the result of evil spirits entering your body, although you do not know why they would do such a thing. Perhaps the spirits are especially malignant. Perhaps you have somehow lowered your defenses. Perhaps the cause is yet another of life's unknowns and is not worth pursuing. Regardless, it is only natural that you turn to tobacco for relief, calling on what you consider to be the good spirits. And it is natural that your priest, or a medicine man with priestly powers, be the person to administer the plant. Joining you in this practice are Aztec and Toltec, Iroquois and Seminole, Algonquin and Micmac. In fact, among virtually all the tribes of the Americas, North and South, tobacco is as important an ingredient in medicine as it is in theology.

One of the first Europeans to note this was the Portuguese explorer Pedro Alvarez Cabral, at the turn of the sixteenth century. In a document not published until some years later, he wrote of the natives he had seen in his American travels:

> They have many odoriferous and medicinal herbs different from ours; among them is one we call fumo (i.e., tobacco) which some call Betum and I will call the holy herb, because of its powerful virtue in wondrous ways, of which I have had experience, principally in desperate cases: for ulcerated abscesses, fistulas, sores, inveterate polyps and many other ailments.

The Englishman Thomas Hariot thought he understood how tobacco worked as medicine. "It openeth all the pores and passages of the body," he wrote at the time, and therefore the native American tribes "know not many grievous diseases wherewithal wee in England are oftentimes afflicted." But it was not just a matter of opening pores. "Unlike herbal remedies, which were infused in water or wine," reports the modern historian Giles Milton, "tobacco was inhaled directly into the lungs, a novel procedure that was said to bring immediate relief."

Another white man, the Spaniard Gonzalo Fernandez de Ovieda ya Valdes, had heard of tobacco use among natives before his journey to the New World. He expected to witness it and was eager to note its effects. But he did not realize that Europeans themselves were starting to smoke when they crossed the ocean—a few of them, at least—and for a reason he would never have guessed. "I am aware that Christians have already adopted the habit," he stated, having crossed the ocean and begun to look around, "especially those who have contracted syphilis, for they say that in the state of ecstasy caused by the smoke, they no longer feel their pain."

The priests and medicine men swore by their methods, despite having no real training in the healing arts. They preferred tobacco to all other nostrums of the time, and prescribed it for as wide a range of disturbances as charlatans of the nineteenth century prescribed their petroleum-based patent medicines. They found tobacco "the most sovereign and precious weed that ever the earth tendered to the use of man." Their patients, by and large, agreed.

Perhaps the first malady to be treated with smoke was fatigue. After a day at his labors, hunting or farming or erecting a structure of some sort, a man might be so weary that he did not think he could go on. But if he fired up his pipe, took a few puffs, and reacted to the changes in his blood vessels and pulse rate and skin temperature, he would feel better in a matter of minutes. Or different, at least—and that was a passable imitation of better. This respite from exhaustion, however brief, would seem a return of vigor. The man might even have another pipe after dinner, this time to aid his digestion, for he believed that "if one but swallow a few mouthfuls [of smoke], even though he should eat a whole sheep, he would feel no sense of fullness whatever."

Inevitably, the quick fix of tobacco was taken for a lasting balm and thus was recommended for more serious ailments. Some tribes, for instance, thought it would cure a fellow of delirium. If he was *so* delirious that he could not work the pipe himself, the medicine man would light his own pipe and exhale the smoke into his afflicted mate's nostrils. This would continue until the victim either regained his senses or lost them altogether. In other tribes, the medicine man would discharge smoke onto whatever body part troubled his patient: a few puffs over the forehead for a fever, a few over the stomach for a cramp, a few over the thigh for a sore muscle. The mere touch of the vapors, warm and wispy as they floated across exposed skin, would surely bring relief.

More bizarrely, the Aztecs and Incas were among those who practiced the rectal application of tobacco smoke. Where they got the idea, no one can say. Why anyone would endure such a treatment is an even greater puzzle. Nor can he who did the applying have been eager for the business. The Jivaros of eastern Ecuador even tended to their children in this manner, rolling them over onto their sides and anally inserting a syringe that was made of a hen's bladder.

To blow smoke up one's ass. Today it means to compliment in a crude and obvious manner; in the past it meant to cure in a manner even more crude.

As civilization advanced, the use of tobacco enemas declined. The latter, in fact, seems almost a prerequisite to the former. But for some reason, tobacco enemas made a comeback in eighteenth century Europe, where they were utilized "to resuscitate people in a state of suspended animation, or apparently drowned persons." Sometimes, instead of a syringe, a bellows was inserted in the patient's posterior opening, thereby increasing the horsepower of the therapy. In France, a certain Doctor Buc'hoz came up with a variation on the theme that was even more horrifying, as he expelled smoke into the vaginas of women who came to him for hysteria. To the modern observer, the method seems more likely to cause the condition than make it go away.

And in the United States in 1893, a woman named Adelaide Hollingsworth published *The Columbia Cook Book*, which, despite its title, was no mere collection of recipes. Rather, Hollingsworth intended the volume as an aid to women in all their household duties. If she happened to poison herself, Hollingsworth recommended that she induce vomiting. The best way to do it: "Injections of tobacco smoke into the anus through a pipe stem."

But in early times, at least, there were more medicinal uses for the leaves of tobacco than for the smoke. Some natives would grind them up and mix them with water, making a paste of them and rubbing it on a woman's abdomen to prevent miscarriage. Others would create a suppository for malaria, a poultice for diarrhea, or an ointment for sore muscles, in this case combining the leaf paste with ground insects, the more venomous the better. And there were emetics for flatulence, dentifrices for toothaches or bleeding gums, and salves with lime an added ingredient, for various kinds of sores. One could pass a leaf over his body, like a talisman, "to drive out the fright, *aigre*, or foreign matters"; one could fry a leaf in butter and apply it to a cut for clean, quick succor. There were also tribes of early Americans who would toast leaves and stuff them under their armpits to relieve pains of an unspecified nature.

If a person had a headache, he might sniff tobacco in powdered form. If he had a cold, he might shred some leaves and insert them in his nose, packing them loosely enough to permit breathing but tightly enough so that at least a few would remain in place after a sneeze. If he had a neck ache, he might lie with his neck on a large green leaf, or if he had to be up and about, he would wrap the leaf around his throat, securing it with a vine or piece of cloth.

"The remedy for snakebites" in the New World at this time "is the sucking of the place where it was bitten and to make an incision and place a thin, transparent piece of cloth made of maguey over it, expose the bite to fire, to warm it and to rub it with ground tobacco." On occasion, a tribe would assign some of its braver members to deposit tobacco leaves in snake-infested areas: on the ground, around rock formations, on the leaves and branches of trees and other plants. The hope was that, as the reptiles slithered around looking for food, they would gobble up the tobacco and become too intoxicated to bite. It is not clear that this ever worked.

To the Chorti Maya, as to natives of most tribes, the medicine man was a figure of unquestioned legitimacy. The stranger his notions seemed, the more they were viewed as a form of expertise beyond the attainment of others. Therefore, he heard no complaints when, dealing with seizures, he began chewing tobacco until he had worked up a good lather, then spit the juice all over his patients, several spurts' worth, letting it land where it might and soak into the skin. "If the illness is especially severe," Francis Robicsek tells us, the medicine man "sprays his saliva-mixed tobacco in the form of a cross from head to crotch and shoulder to shoulder."

Other disabilities thought to be curable by tobacco in one form or another, at one time or another, were ague, asthma, "bad humours," blood poisoning, boils, bowel rumblings, bruises, carbuncles, catarrh, chilblains, colic, constipation, convulsions, cysts, dirty teeth, earaches, "excessive or superfluous phlegm," "falling nailes of the finger," gonorrhea, halitosis, herpes, hoarseness, hydrophobia, "internal disorder," itchy eyes, joint pains, kidney stones, lupus, paralysis, poor vision, pulled muscles, rheums, scabs, scurf, skin eruptions, sore throats, stings, swellings, tension, tetanus, "tight bowels," warts, "windy griefs of the breast," and worms.

If a person had the gout, he might dip his feet into a tobacco-tea foot bath, "using ... leaves that had first been left in a ditch where ants could walk on them, followed by 'serpentine rabbits' ground into a powder, a small white or red stone, a yellow flint and 'the flesh and excrement of a fox which you must burn to a crisp.'"

Even poor eyesight, it was believed, could be improved by tobacco, with the plant's juice sometimes being administered as eye drops. Hunger, too, would inevitably yield to tobacco's divinely bestowed powers. All a person had to do was combine the leaf with ground shells and form the resulting compound into pills: "one could live up to a week on these alone without any other food." Tobacco was Advil and Afrin and Robitussin, Visine and Listerine and Ex-Lax, hydrocortisone cream and analgesic balm and hot pack and ice bag and a snack-size bag of Doritos—all of these, all rolled into one.

A FEW TRIBES favored preventive medicine. Those of the Great Lakes, for example, were especially cautious. If they saw a large boulder or a misshapen tree trunk in front of them in their wanderings, they would stop immediately, utter a prayer or two, and then sprinkle some tobacco leaves around them. This, they had determined, would ward off the malicious spirits who lived in the rock or tree before they could escape and spread their infections. For the same reason, tribes around Virginia and the Chesapeake cast powdered tobacco over the waters, while others dispersed it into their fires, sending the flames crackling into the night, producing a hot and shimmering protective shield. In South America, there were tribes rubbing tobacco juice onto their skin, confident that it would kill lice and other bugs.

Some natives even went so far as to steep tobacco leaves, make a hot beverage of them, and give it to a woman as she was bringing new life into the world.

Whenever a pregnant woman in great pain
Cannot part from the child
And is unable to give birth to it,
Then she would drink the water of the weed,
And in no time she will bring forth
The fruit of her womb.

But on occasion, the woman's pain was unbearable. She could not get the beverage down. A midwife might come to the rescue. She would tilt the mother's head back and pour the steamy hot liquid into her nostrils, allowing it to gurgle through the sinuses, slosh against the roof of the mouth, and gush down her throat the long, hard way. The mother would react violently, coughing and spitting and gagging, thrashing at the midwife to push her away. Undaunted, the midwife would grab on to her patient and repeat her steps. The mother would repeat hers. On and on this would go as the mother, all the while, heaved and pushed and grunted. After this, she must have looked at the actual rigors of childbirth as little more than a minor inconvenience.

Like smoke, this tobacco tea was also, on occasion, taken rectally, "where it was introduced in the form of a clyster, using a hollow length of cane or bone, or with a bulb made out of animal skin and a bone or reed nozzle." In none of my research could I discover what bodily malfunctions were treated in this manner or how well they responded.

THE LEAF COULD do so many things. And a man could create it himself—or, at least, bring it into being: planting seeds and watering them and watching them grow, harvesting the crop when it was ready, enjoying it when he wished and as he wished. He did not just avail himself of tobacco; he ordered its various duties.

Not so with smoke. Smoke was a different matter entirely. Smoke cannot be defined, can barely be described, and has an almost mystical lineage. It is, after all, the product of fire, the earth's first miracle, which is in turn a little piece of the sun, the first miracle that early man saw in the vast spaces above him.

Fire is more vivid than smoke, more colorful and menacing. The eye is drawn to it as the breath races and the heart beats faster. There is a sense of unpredictability, of danger. But smoke is the freer spirit. Nothing controls it, no one is its master, perhaps not even the gods. It can escape from the confines of fire to ascend on its own, and so much farther does it travel than the flames beneath it that on occasion the flames seem to have no other purpose than to provide a launch. Smoke is here and then it is there and then it is out of sight altogether, until the next streams come along. You can touch smoke and still be empty-handed; you can feel it brush against you and not be sure you have made contact. It never looks the same from one instant to the next, never assumes the same shape or shadings or takes the same path twice. Smoke is the embodiment of a whim, a material for dreams, for yearning. It is as insubstantial as a substance can be and still be real. In a world of solids and liquids, and not all of *them* comprehensible to men and women of the ancient world, smoke could not help but stand out—and so it was that the native tribes of the Americas were not content to save tobacco for ceremony and sickness. Nor were they willing to have its uses determined solely by priests and medicine men. Smoke was something that every person wanted for himself, and for that reason it would in time be freed from specialization, made an integral part of everyday life.

People turned to tobacco as they turned to food and water. They turned to it morning, noon, and night, before eating, after eating, and without regard to meals at all. They turned to it in their dwellings, in their workplaces, in their public areas. They smoked alone and in the company of others, but the latter was the more important setting. A native might not be sure that his vapors went all the way up to heaven, but he could see them reach his companions, wind around them, a literal bonding, just as the sinuous trails of smoke produced by others would wind back to him, completing the circle of amity.

In the societies that had such things as courts, smoking became a favorite royal activity. Pipes were passed around during banquets; favored guests might be rewarded for their attendance with a package of leaf to go. The Aztecs taught that the goddess Cihuacoahuatl was actually made of tobacco, and the emperor Montezuma worshipped no deity with greater passion. When he feasted, servants set before him "three painted and gilded tubes containing liquidambar mixed with a certain plant they call *tobacco*; and after they had sung and danced for him, and the table was cleared, he took the smoke of one of those tubes, and little by little with it he fell asleep."

Awake, the Aztecs and other native tribes made tobacco into a rite of government. Members of a tribal council would light their pipes before a meeting officially began and keep them going until they had addressed the last item on the agenda, believing that this enabled them to work more efficiently and for longer hours. Further, as the novelist William Makepeace Thackeray would in time write, tobacco helped people express themselves better, giving them

> great physical advantages in conversation. You may stop talking if you like, but the breaks of silence never seem disagreeable, being filled up by the puffing of the smoke; hence there is no awkwardness in resuming the conversation, no straining for effect—sentiments are delivered in a grave, easy manner.

The natives believed, in other words, that the more they smoked, the better they legislated, which meant that the air at their meetings would often grow thick and gray, the faces of the members becoming as indistinct to one another as the countenances of a mirage. In fact, according to at least one educated guess, the Quiche Maya "may have originated the 'smoke-filled room' of politics, since their councils were illuminated by fat-pine torches and accompanied by fat cigars."

So it was not just pipes. The first Americans were also smoking a forerunner of the modern cigar, and on some occasions it was *so* fat that it could not be employed without assistance. "Some tribes," it has been written, "developed special cigar supports, resembling giant tuning forks, which could be held in the hand, or whose sharp end could be stuck in the ground to support these monsters."

It is not certain whether the pipe or the cigar came first, although most accounts of tobacco's early days concentrate on the former, and the earliest known picture of a smoker, carved into the wall of a Mayan temple in the ancient Mexican village of Palenque, shows a pipe in the mouth of a priest, his head encircled by ornate patterns of smoke that drift around him and wend their way skyward. If the bowl had not been shaped like a god, it might take the form of a bird, a creature that could follow the smoke, ride along its paths, and even exceed them.

But a cigar was easier to construct and operate than a pipe. At one point, it seems to have become the item of choice for most of the tribes of the Americas. Assuming it was not one of the "monsters," it fit into the hand as nicely as did a pipe, with each lending itself to gestures of conviviality, bursts of expressiveness. Which is to say that both the cigar and the pipe were genuine trappings of civilization and as such played a more important role in the progress of various native cultures than many historians acknowledge. Serving as totems no less than tools, pipes and cigars stood for behaviors of which their users were proud. Over the years, though, they took on very different associations.

The cigar represented the spear with which a man battled another man or a beast. He who smoked one thus advertised himself as a more robust sort than the pipe smoker, and not just because of the symbolism. He had also chosen the stronger tasting of the two tobacco mediums, as well as the one with the more potent aroma. A man announced his presence with a cigar even when he was some distance away. He let people know that he was not just a member of the tribe but an individual to whom attention must be paid.

The pipe, however, had more communal implications, to a degree because it demanded so much of the smoker:

> The tobacco must not be too dry, lest it bite, nor too moist, lest it fail to hold fire; the pipe must be caked, so as not to yield a raw taste, yet not soggy, so as not to impede the draw. Its maintenance requires copious pockets and large supplies of patience, matches and leisure.

The pipe stood for peace and cooperation, most notably among the natives of the Mississippi Valley and Great Lakes. It was the nature of the thing; it encouraged deliberation. When tempers flared at the tribal assemblies, smoking a pipe was like counting to ten, if not higher. It gave a person time to cool off, to devise a reasoned response and then attend once more to business.

With the passage of years, the peace pipe was taken up by more and more tribes. When Meriwether Lewis explored the Louisiana Territory with William Clark early in the nineteenth century, he offered a pipe to some of the natives he met along the way, thinking of it as a casual gesture. Only later did he learn that, to the recipients, the gift was a pledge of honor on both their parts. For the natives in particular, Lewis wrote, smoking a pipe with the white man showed a commitment toward harmony and even, if possible, brotherhood. It "is as much to say," Lewis went on, "that they wish they may always go bearfoot [*sic*] if they are not sincere; a pretty heavy penalty if they are to march through the plains of their country."

A myth of the Oglala Sioux has it that the peace pipe was a gift from a beautiful, long-haired woman who came out of a cloud one day wearing "a fine white buckskin dress." She handed the pipe to a chief and let him know that it was a gift of special significance. "Behold!" she said. "With this you shall multiply and be a good nation. Nothing but good shall come from it."

In his "Song of Hiawatha," the poet Henry Wadsworth Longfellow invents a myth of his own, telling of a chief named Gitche Manito, who broke a piece of red stone from a quarry and sculpted it into a bowl, attaching a long reed from a nearby river as the stem. He "filled the pipe with the bark of willow," and then, as he stood upon a mountain and looked out at the lands beneath him,

> *Gitche Manito, the mighty,*
> *Smoked the calumet, the Peace-Pipe,*
> *As a signal to the nations.*
>   *And the smoke rose slowly, slowly,*

Through the tranquil air of morning,
First a single line of darkness,
Then a denser, bluer vapor,
Then a snow-white cloud unfolding,
Like the tree-tops of the forest,
Ever rising, rising, rising,
Till it touched the top of heaven,
Ever rising, rising, rising,
Till it touched the top of heaven,
Till it broke against the heaven,
And rolled outward all around it.

Encountering the pipe in the seventeenth century, the French missionary Jacques Cartier, who explored the Mississippi River to the Gulf of Mexico, was initially mystified. He stood with some natives and watched them "suck so long that they fill their bodies full of smoke, till that it cometh out of their mouth and nostrils, as out of the tunnel of a chimney." Yet Cartier was intrigued, curious. He and his mates "tried a few drags from the lighted pipe, but found the smoke too hot to bear, 'like powdered pepper.'" At least for the time being, they smoked no more.

It was clear to Cartier, though, that the practice so strange to him had great meaning to those who engaged in it:

> There is nothing more mysterious or respected than the pipe. ... Less honor is paid to the crowns and scepters of kings. It seems to be the god of peace and war, the arbiter of life and death. It has to be but carried on one's person, and displayed, to enable one to walk through the midst of enemies, who, in the hottest of fights, lay down their arms when it is shown.

Such were the encomiums for tobacco in days long past; such was the weight of hopes and promises that rested on leaves so fragile. It was from the start a commodity that could be adapted to many ends, that could be dressed up in so many convincing disguises that it would take centuries for human beings to strip away all the layers and work their way down to the bitter, uncompromising, and often fatal truth.

ONE

The Old World

HE HAD SUCH HIGH HOPES. Visionaries always do. Christopher Columbus, this "obscure but ambitious mariner," this dreamer of large dreams who was "considered a little touched in the head," this man who had vowed to sail to the west to arrive in the east, was about to cast off, to make good on his word or perish in the attempt. He would not prove the Earth was round; most people of the time already knew it, or suspected it, if for no other reason than that they could stand on the shore and, looking out, watch a ship gradually disappear at the horizon. Actually, Columbus thought the Earth had "the shape of a pear ... or that it is as if one had a very round ball, on part of which something like a woman's teat were placed."

What Columbus intended was to find a short route to the Indies and, in particular, to China, which, in the words of Samuel Eliot Morison, "cast a spell over the European imagination in the fifteenth century. These were lands of vast wealth in gold, silver and precious stones, in silk and fine cotton, in spices, drugs and perfumes."

Columbus would locate these treasures. He would fill up his chests with them and load them onto his ships and bring them back to Europe, where he would receive the hero's welcome he believed was his due and accept it with the magnanimity he liked to think was his nature. In the process, he would create new trade routes and perhaps make discoveries that would lead to the redrawing of maps, which is to say, to the reconfiguring of reality. He would justify the faith of the Spanish monarchs, Ferdinand and Isabella, who had sponsored him, and they in turn would reward him with titles and security and possibly even a share of the booty, a finder's fee.

At the same time, he would not be a slave to his magnanimity. He would thumb his nose at those who had said his mission was madness and from this he would take special satisfaction. His name, he dared to imagine, would be known around the globe, and all that he touched, each place he visited, would be changed evermore because of his appearance.

As things turned out, Columbus was justified in his hopes. He earned the praise of Spain's king and queen and all manner of others, noblemen and peasants alike. He won his titles, one of which was Admiral of the Ocean Sea. He gave work to several generations of cartographers, and those who had scoffed at him were forced to retreat in shame, or at least hold their tongues.

But his specific accomplishments, as significant as they were, proved to be far different from what he had foreseen. Among them, though he did not know it at the time and in fact never would, was the introduction to Europe of a fragrant, brown, New World plant, previously unknown to the Old. Some historians believe Columbus brought it back with him, that it was one of the assembled treasures despite its humble appearance. Others, who seem the minority, insist that Columbus merely reported on the plant, telling some peculiar tales of native consumption and inspiring others to locate and transport it. Regardless, the plant would prove, in the words of the Spanish historian Manrico Obregon, to be "as valuable as gold, and perhaps as harmful."

HE SET OUT from Palos, Spain on August 3, 1492, with three ships, two of which would make it back, and a crew of ninety. They left before sunrise, although a few orange rays had begun to bleed into the eastern sky. As Columbus himself put it in the first of his journal entries, he and his men sailed down the Rio Tino toward the open sea, proceeding "south with a strong, veering wind until sunset."

But it was a journal entry of October 15, more than two months later, after Columbus had put the Atlantic Ocean behind him, reached the Caribbean, and begun his futile search for signs of the Orient, that catches our eye:

> Halfway between these two islands, Santa Maria and the larger one which I am calling Fernandina, we found a man alone in an almadia [a kind of canoe] making the same crossing as ourselves. He had a piece of bread as big as his fist, a calabash of water, a piece of red earth, powdered and kneaded, and a few dried leaves which must be something of importance to these people, because they brought me some in San Salvador.

The people were Tainos, members of the Arawak family of native Americans. Peter Martyr, who has been called the first chronicler of the New World, described them as farmers, weavers, and pottery makers, handsome people of kind disposition who "seem to live in that golden world of which old writers speak so much, wherein men lived simply and innocently without enforcement of laws, without quarreling judges and libels, content only to satisfy nature."

Columbus was equally fond of the Tainos. "They invite you to share anything that they possess," he told his journal, "and show as much love as if their hearts went with it."

And the dried leaves, which seemed to "be something of importance"? They were tobacco, and as Columbus wrote, he had received some of them earlier in his journey, in San Salvador, on which occasion they had been accompanied by several varieties of fruit. He and his men were confused by the offering, not knowing what to make of it but not wanting to hurt anyone's feelings. We may imagine that they hemmed and hawed, stalled for time, traded unknowing glances with one another. Finally, sensing that the natives were becoming annoyed with their indecision, they did the only thing that seemed reasonable: threw away the tobacco, ate the fruit. At which point, the puzzled expressions must have belonged to the tribesmen.

It was not until another month had passed and Columbus and his fleet had traveled still farther west that they first saw the use to which the natives put the leaves—and it was a startling sight to them, perhaps the strangest of all on the far side of the world. It was also an important, if largely unrecognized, moment in cultural history, the first step on tobacco's long and tortuous journey from primitive custom to emblem of established society to badge of opprobrium.

Certain now that he had arrived in China, although for what reason no one can say, as he had not encountered so much as a fiber of corroborating evidence, Columbus dispatched two of his men, along with a party of native guides, to look up the nation's ruler, the Great Khan. The Europeans were Luis de Torres, an Arabic scholar, and the able seaman Rodrigo de Xerez. Their instructions were simple: proceed to the khan's palace, introduce themselves, present him with some beads and other trinkets, and tell him that Christopher Columbus, representing a great people from very far away, would like an audience. They were to make sure that the Chinese understood it was commerce the visitors had on their minds, not conquest. They were, in other words, to put the mighty ruler at ease.

Of course, they did not. They never found him. What de Torres and de Xerez did find is recounted many years later, as vividly as if he had actually been there, by the Spanish writer Bartoleme de las Casas.

> The two Christians met many men and women who were carrying glowing coal in their hands, as well as good-smelling herbs. They were dried plants, like small muskets made of paper that children play with during

> the Easter festivities. They set one end on fire and inhaled and drank the smoke on the other. It is said that in this way they become sleepy and drunk, but also that they got rid of their tiredness. The people called these small muskets *tobacco.*

The two Christians—one of whom, de Torres, was probably a Jew—did not record their impressions, but it is certain that they had never seen anything like this before. Man, it has been observed, is the only animal who takes smoke into his body for pleasure, but it had not been observed back in Columbus's time, and even if it had been, there is no way that de Torres and de Xerez would have believed it. As Iain Gately writes, "No one smoked anything in Europe. They burned things to produce sweet smells, to sniff, but not to inhale. Smoke was for dispersal, not consumption."

In fact, far from being a source of pleasure, the smoke-drinking might have seemed to the Europeans a form of punishment—self-inflicted, but punishment nonetheless. It may be that they had nothing with which to compare it except for tales of fire-breathing monsters they had heard as children, legends of anguish and destruction that parents told to children, or rulers to subjects, to frighten them into obedience. For de Torres and de Xerez, this can hardly have been reassuring.

No one knows how the natives consumed their tobacco that day. In all likelihood, they inhaled the smoke conventionally, through their mouths, but it is also possible that they took it in nasally. The latter was a practice of many tribes at the time and would continue to be, in some parts of the New World, for years to come. A smoker would use a hollow, Y-shaped cane, similar in appearance to a slingshot but not as wide across the top. He would pack the top ends into his nostrils and place the bottom at the tip of his musket, breathing in the smoke through this device, working up a holy buzz.

Nor is it known precisely how the natives exhaled. Because the white man was so extraordinary in appearance, because his manner and dress were so different from anything that the inhabitants of the New World had seen before, many tribes perceived him as a god, or at least an empyreal agent. If this is true of the men who encountered de Torres and de Xerez, they might have blown their smoke directly onto the two Europeans, enveloping them in emissions as a means of worshiping them—these particular natives, like the Mayas and others before them, making contact with supernatural beings through the wispy offices of tobacco.

Later voyagers to America were beclouded frequently. Take the Frenchman Nicholas Perrot, who, about two-thirds of the way through the seventeenth century, found himself standing before some unidentified tribesmen in an unknown outpost of the continent. The natives clearly adored him, seeming to regard his presence as the answer to their prayers. Yielding to their wishes, Perrot allowed them to exhale into his face from a distance of but a few inches, a chorus of idolaters spewing their emissions in unison. It was "the greatest honor they could render him; he saw himself smoked like meat."

The Frenchman took it well, all things considered. He nodded, silently departed, and went on with his day.

De Torres and de Xerez reported back to Columbus. There is no account of what they said, or of their skipper's initial reaction. Did he believe the tale his men had told? Did he think something in the New World had bewitched them, causing them to see what did not exist? Did he think they were playing a joke on him, perhaps testing his sanity after the long voyage? Whatever he concluded, by the time Columbus got around to his journal that night he had put the episode into prosaic perspective, relating it is just anther event on a typically eventful day in uncharted territory. The date was November 6, 1492:

> My two men met many people crossing their path to reach their villages, men and women, carrying in their hand a burning brand and herbs which they use to produce fragrant smoke.

The key to the entry is the word "fragrant." Columbus and de Torres and de Xerez could only assume, after much reflection, that smoking was a means the Indians had developed to perfume themselves. No other theory came to the European mind, even though the scent almost surely did not appeal to them. One more strange custom in one more strange corner of the earth.

Columbus kept on with his explorations. Having already reached the Bay of Bariay off Cuba, he continued to send scouting parties ashore, and they explored the island and several smaller islands nearby. All the while he looked for signs of the Orient; all the while he found dark-skinned men and women with humble ways and muskets of tobacco, and no one who had even heard of the Great Khan.

Before setting sail for home, Columbus made his way along the northern coast of Hispaniola. Here, as elsewhere, he came across Indians at sea, rowing alongside his larger vessels in their dugouts, gaping at him and his men, smoking, always smoking. The more the white men saw of the custom, the less grotesque it seemed, and the more willing they became to try it themselves. Perhaps, on the voyage back to the Old World, there were moments when tiny clouds of smoke floated furtively over the decks of European ships, vanishing after a few seconds in the damp, salty air.

If Columbus actually did carry tobacco with him when he arrived in Spain in March 1493, he almost certainly meant it to be a souvenir, a token of a different way of life, a curiosity, in the same way that a tourist of the present will not bring home anything practical, but the most unusual handicrafts he can find from the most remote stops on his vacation. Columbus never imagined that tobacco would appeal to his sophisticated countrymen and did not live long enough to see it happen. And he probably did not know of the fate that befell his old colleague, the able seaman.

Rodrigo de Xerez packed some tobacco for the return voyage, perhaps in secret, and by the time he set foot again in his hometown of Ayamonte in southwestern Spain, he was hooked. Overcoming his initial resistance, he had decided to sample a musket, doing as the natives did, and before long he was smoking as much as they were and seeming to derive the same amount of enjoyment, if not the same kind of spiritual release. As the only man in town, and probably on the entire European landmass, so addicted, he was as offensive a sight to his friends and neighbors as the Tainos had once been to him. The Spaniards could not understand what had happened to dear old Rodrigo on that voyage to the Indies. Perhaps he had become ill. Perhaps he had lost his mind. Perhaps, for some reason, he was trying to shock.

No, they decided after a time, it was even worse. De Xerez had become possessed by the devil on his journey. The New World was a precinct of hell, and the smoke that he now expelled came from the very fires of perdition. The citizens of Ayamonte discussed the matter among themselves and decided there was but one course to take: report the evildoer to the Inquisition, that most ruthless of crusades for religious orthodoxy. In the long run, they persuaded themselves, they would not be punishing de Xerez so much as setting him back on the right path. He would thank them. He would realize the wisdom of being an example to others. He would also stop releasing those vile and noxious clouds from his nose and mouth.

The Inquisitors, as was their way, showed little mercy. They stripped de Xerez of his material goods, confiscated his land, and threw him into jail—for three years, according to one account; for seven, according to another. Thus was the sailor Rodrigo de Xerez, in more ways than one, the Old World's first victim of the newly arrived, instantly disdained plant called tobacco.

THE DECADES THAT followed were the golden age of exploration. The world we know today began to take shape in the sixteenth century, as places were discovered and named and new patterns of trade and cultural exchange came into being. New ideas were advanced, new products introduced; newness itself became an expectation rather than a novelty. Old customs fell into disuse, old institutions reformed themselves or faded away, populations shifted. Wars were fought that lasted for years, and treaties were signed that fell apart in moments. There was colonization and exploitation, perplexity and discontent, thievery and barbarity, and an unsettled quality to life that seemed to astonish men and women as they went here and there and back again more often and more rapidly than ever before. The cartographers were not the only ones kept busy. So, too, were all who tried to stay abreast of the times, to hold on to what had once been, to look at their lives according to long-accepted standards.

And tobacco was a part of it all, writes Daniel Klein—an important part. People turned to it "as a drug for easing the anxiety arising from the shock of successive assaults on old certainties and the prospect of greater unknowns." At no time in the past had human beings faced so extraordinary a problem.

Juan de Grijalva discovered Mexico and Pascuel de Andagoya discovered Peru and Amerigo Vespucci and Alonso de Ojeda discovered the mouth of the Amazon in South America. Giovanni de Verrazano came upon New York Bay and followed the Hudson River to the north, and the Cabots, father and son, landed on Cape Breton Island off Nova Scotia and moved inland from there. Rodrigo de Bastides went to Panama, and Vasco Nunez de Balboa crossed the Panamanian isthmus and cast his eyes on the Pacific. Pedro Alvarez Cabral claimed Brazil for Portugal, and Jacques Cartier ended up in Labrador and Juan Ponce de Leon in Florida, while Ferdinand Magellan circumnavigated all of it, the entire globe, a literal round trip. Vasco da Gama, who had previously found the water route to the Indies that had been Columbus's dream, circled the Cape of Good Hope and eventually returned to India as Portuguese viceroy.

As for Columbus himself, he had proved a better mariner than a people person. He made three more trips to the New World, all of them more troubled than the first. Eventually, due to the unhappiness of his crews and discontent in the colonies he had founded, King Ferdinand stripped him of at least one of his titles, as well as his financial stake in New World trade. Columbus pleaded for their return. More than once he was seen trekking by mule to the royal palace—a woebegone sight, this proud man reduced to such a pathetic mission on such an ignoble beast—hoping the king would see him and relent. He did neither. Plagued in the final months of his life by arthritis no less than disrepute, Columbus died in humble surroundings on May 20, 1506.

And when they returned to their home ports, these men of adventure who had finished for the time being with their travels, they carried with them as much of America as would fit into their ships, holds bulging with statues, trinkets, playthings, foodstuffs, objects of worship, articles of clothing, the odd piece of furniture, a few gold nuggets, a few pieces of jewelry, a few animals not native to Europe, and even a few of the men and women and children who wore or ate or made other use of the preceding. And, of course, they returned with tobacco, the New World's biggest surprise, introducing it to the Old and then watching and wondering at what they had wrought.

TOBACCO WAS SLOW in getting to England, probably not making an appearance until midway through the sixteenth century, if not a few years later. Sir John Hawkins, a stalwart sea captain and vicious slave trader, is credited with the first deliveries of leaf to his homeland, the result of his raids of tribal settlements along the coast of Florida. As Hawkins put it himself, "The Floridians when they travell have a kinde of herbe dried, who with a cane and an earthen cap in the end, with fire, and the dried herbe put together, doe sucke thorow the cane the smoke thereof, which smoke satisfieth their hunger, and therwith they live foure or five dayes without meat or drinke."

Sir Francis Drake, known to some as the "Master Thief of the unknown world," and Sir Martin Frobisher were among others who forcibly removed tobacco from colonial outbacks and native peoples, as well as from ships along the trade routes. They, too, transported it back to England. But Hawkins beat them to it, and so it was he who became memorialized in verse:

Up comes brave Hawkins on the beach;
"Shiver my hull!" he cries,
"What's these here games, my merry men?"
And then, "Why blame my eyes!
Here's one as chaws, and one as snuffs,
And t' other of the three
Is smoking like a chimbley-pot—
They've found out Tobac-kee!"

But Hawkins, Drake, and Frobisher merely toted the goods. For tobacco to be accepted in England as thoroughly as it was, for it to become so entrenched that it would prove crucial to the economy at the same time that it poisoned relations with the empire's later American colonies, tobacco needed someone to beat its drums, tout its powers, its potential for bliss. It needed, in other words, a press agent, and that was the role eagerly assumed by Sir Walter Raleigh, whose name is correctly spelled without the "i" but more commonly known with it.

Raleigh was one of the outstanding figures of his time, a hard man to categorize and an even harder one to keep up with. As a warrior, he distinguished himself both on land and at sea. As an explorer, he made two courageous but unsuccessful attempts to settle North America, both at Roanoke Island; the second was the occasion of the first English birth in the New World, the child named Virginia Dare. Raleigh helped Edmund Spenser publish his best-known poem, *The Fairie Queene*, and was responsible, at least in part, for introducing the potato to Ireland, where it turned into not only the most important crop but almost the entire foundation of life. He died a dramatic, even theatrical death, in which tobacco not only played a part but served as an indispensable prop.

The essence of the man, though, is not to be found in such detail. Rather, we find it in the way he spoke, his tales both captivating and energetically told. We find it in his winning smile and formidable will and his quick, sometimes cold, sense of humor. A dedicated social climber, Raleigh had all the superficial graces and a few of the more profound ones. His charm affected men and women equally, usually attracting them, although sometimes putting them off.

His physical being advertised the charm. At six feet tall, he towered above most Englishmen of the period, the average of whom was about half a head shorter. His skin was swarthier than theirs, almost as if he had New World kin, and his face stood out all the more because of a sharp nose and dark, pointed beard. As for the hair on his head, it was long and curled and sometimes perfumed. He could be menacing if he wanted to. Those who did not know him well were advised to stay alert in his presence.

But he did not look menacing. Rather, he appeared every inch the dandy. "His customary cartwheel ruff was his most extravagant gesture to foppishness," writes the historian Giles Milton, "spreading peacocklike from his neck in dentilated lace. It was a perfect complement to his satin pinked vest and gauche doublet cut from finely flowered velvet and embroidered with pearls." He also wore a pearl in one ear and rings on most of his fingers and "a dagger with a jeweled pommel." He was impossible to miss in a crowd of any size, of any constituency.

What made him so effective a spokesman for tobacco, though, was his relationship with the queen, Elizabeth I, a woman of poise and naturally regal bearing who also had a dismissive manner at times, and meager reserves of patience. She was a vain woman, not entirely trustworthy, tight with the royal purse strings and unacquainted with religious conviction. Yet there was something about Elizabeth that engaged her subjects, earning their admiration as much as their obeisance. Perhaps it was merely a gift for masking her flaws.

Perhaps, more than that, it was a genuine affection for her subjects. Near the end of her life, her strength fading, she addressed them warmly, saying that "though ye have had, and may have, many princes more mighty and wise sitting in this seat, yet you never had, or shall have, any that will be more careful and loving." It might well have been true. Elizabeth I was a perfect fit for her station in life and presided over a glorious time in her native country. She was, to virtually one and all, "good Queen Bess."

Raleigh met her when he was twenty-six and she was in her late forties. He fascinated her from the start. He, in turn, found her to be "someone who shared his insatiable curiosity; and also a woman, barred by her sex from activity and adventures in a wider world, who loved to experience vicariously the life of a thinker who was also a man of action."

But if she gained vicariously from him, he gained tangibly from her. She appointed him captain of her guards and vice-admiral of the West and, later, granted him a knighthood. She also granted monopolies for the sale of Cornish tin and sweet wines, as well as a license to export woolen broadcloth, all of which proved exceedingly lucrative. She presented him with an estate of 12,000 acres in Ireland and a place to call his own in London, known as Durham House. No less important for a social climber, Raleigh found himself in possession of both the queen's time and ear. He was among her most trusted advisers and would remain so until another Elizabeth entered his life, this one a commoner with whom he behaved in what was for the era a most *un*common—and shocking—manner.

Perhaps out of gratitude, perhaps feeling something deeper, Raleigh would write poetry for his queen. On one occasion, he even scratched a few lines of verse onto a palace window. And, on all occasions, he treated her with great solicitude, seeming in her company to have no concern other than her comfort and well-being. Once upon a rainy day, it has been written, Raleigh removed his "new plush cloak" and spread it upon the ground so that Elizabeth could cross "a plashy place" in the road without getting her feet wet. The story is almost certainly apocryphal but worth repeating, as so many have done, for what it tells us about the friendship between monarch and citizen.

But that is all it was, a friendship. Elizabeth and Raleigh were not lovers, not physically. She might have flirted with him and might also have given a wink or a nod or a smile to other members of her court from time to time. But the lady had no lovers, never did or would, and most of the realm knew it. In fact, when Raleigh was trying to make a go of his New World settlements, he gave to one of the colonies the name Virginia, in honor of—or possibly lamenting—the state of Elizabeth's relations with the male of the species over the entire course of her life.

It was inevitable, then, that when Sir Walter Raleigh showed up at court one day with a pipe in his mouth, probably one whose bowl had been whittled into a small bust of Raleigh himself, Elizabeth would listen carefully to his raptures. And raptures they were: He told her that tobacco gave him a feeling like no other he had ever known. He told her it rid him of pain and sickness and would do the same for her in a matter of but a few puffs. He told her that her courtiers and ministers and even some of her ladies were beginning to smoke and were much attracted to it, their initial titillation already turning to a more profound enjoyment. Then he told her she could trust him completely when it came to the strange, foul-looking New World plant. He knew so much about it, he said, that he could even weigh its smoke:

> Always keen for a wager, Elizabeth bet him he could not make good his claim, whereupon Walter called for scales. Pinching some tobacco shreddings from the gilded leather pouch he always carried with him, he carefully weighed the amount he needed to fill his long stemmed pipe, smoked it and then weighed the ashes. Subtracting this second weight from the first he produced the answer that would win his wager, Elizabeth remarking as she paid up that she had heard of men "who turned gold into smoke, but Ralegh was the first who turned smoke into gold."

The queen was easily won over. So were the literati of Elizabethan England, men like Ben Jonson, Christopher Marlowe, John Fletcher, Francis Beaumont, and possibly even William Shakespeare, although he seems never to have written about smoking, either publicly or in private. It all started on the night when Raleigh breezed into the Mermaid Tavern, London's official authorial hangout, with pouches of tobacco and an armful of pipes, which he handed out like a duke dispensing alms. He might also have distributed some of the gadgets that made the pastime of pipe smoking so intricate an endeavor, such as "a metal stopper to press the tobacco into the bowl, a gold or silver pick to cleanse the bowl, a knife to shred tobacco … a scoop for loose tobacco, and whatever else appealed to the playboy as necessary." The latter perhaps included boxes in which a person carried his tobacco and tongs to transport it from box to bowl. In fact, after a time, "the average gallant required so many smoking accessories … that a dedicated manservant was needed to carry them."

Light up, Raleigh told the Mermaid crowd, and light up the men of letters did. Most were impressed. If nothing else, they had found a new affectation, always a blessing to artistic types. Raleigh's own appeal combined with that of the leaf to inspire a set of converts who would soon spread the word, and eloquently—in prose and poetry and via characters on the stage throughout Elizabeth's realm.

Others, though, were a harder sell. Among the people most resistant to the leaf were farmers and miners, fishermen and clerks, artisans and craftsmen, and the various smiths and wrights and label-less laborers who made up the lower classes of the time. They would often stop on the street when a smoker went by, staring at him as if he were an animal never before spotted in these parts. Sometimes they appeared angry, other times frightened, at the bizarre spectacle that the smoker offered. They might turn hurriedly away; they might confront the offender in a defiant manner, questioning his motives or his sanity or both. They might even threaten physical harm, although they probably did not want to get close enough to the tobacco user to land a blow.

Children would run and hide, and possibly from a safe distance call out their derision. To most it was a scandalous form of behavior, all the more so because the smoker always seemed so self-assured in the face of the nonsmoker's consternation, as if he simply could not understand the fuss. According to lore, the first time Walter Raleigh's servant saw his master sitting in his favorite chair, contentedly producing smoke from his pipe, he was so unnerved that he poured a keg of ale over Raleigh's head. Like the tale of the cloak and the puddle, this episode also seems both unlikely and enlightening.

But before long, the queen's blessing, and the increasing use of tobacco by her highly placed subjects, overcame the resistance of the masses. Call it the trickle-down theory of fashion. By 1614, no more than fifty years after it was first unloaded on British docks, tobacco could be purchased at as many as 7,000 establishments in London alone, sometimes for its weight in silver.

Many of the establishments were alehouses, sometimes known as "tabagies," for the English, like Columbus's two Christians, thought that a person took in smoke like a beverage and became similarly intoxicated. It was for this reason that the leaf was sometimes referred to as "sotweed." It was also available at the apothecary's shop, the grocer's, and the chandler's, to name but a few locations. Among innkeepers a kind of competition developed: Raleigh smoked his first pipe in *my* place, one claimed. No, said the proprietor of the inn across the street, he smoked it in *mine. Mine*, shouted the fellow around the corner, even more insistent. In time, the boast was as common as the later one in the United States that "George Washington slept here."

By now the English were puffing at public events, "at bull-baiting, at bear-whipping, in the courtroom, and everywhere else." They were puffing at the Globe Theatre as they watched the plays of Shakespeare and others; they were puffing as they strolled in public parks or took in puppet shows or minstrel performances in the streets. And they were puffing on the way from home to church and again from church to home, although there is no evidence that they saw a connection between smoke and salvation.

They did, however, see a connection between smoke and good health, just as the tribes of North America had seen it before them. After all, says Jerome E. Brooks, "There were supposed to be great curative faculties in all the botanical products of the new-found world, and there was a pathetic need, in lands often ravaged by plagues and endemic diseases, for an unfailing prophylactic herb."

Tobacco was not unfailing. It was not even a prophylactic. But so stubbornly did the English believe in it that, after Elizabeth's time, they even saw it as protection from what was surely London's worst outbreak ever of the plague. Between 1603 and 1665, more than 150,000 people died and many thousands more suffered so greatly before recovering that there were probably times when death seemed to them the more desirable alternative. Outlying districts were also ravaged. Victims who smoked were accused by many of bringing about their ailments by not smoking enough. As for non-smokers, they were urged either to take up the leaf themselves or seek the company of those who did and breathe in the fumes that the others so graciously expelled. Samuel Pepys, that most famous of British diarists, thought it good advice:

> This day [he wrote], much against my will, I did in Drury Lane see two or three houses marked with a red cross upon the doors and "Lord have mercy upon us" writ there; which was a sad sight to me, being the first of the kind that, to my remembrance, I ever saw. It put me into an ill conception of myself and of my smell, so that I was forced to buy some roll tobacco to smell and chaw, which took away the apprehension.

And as the plague was finally disappearing, the lads at Eton who got the worst beatings from the school's taskmasters were not the poor students or the rabble-rousers or the apostates, but the young gentlemen who refused to inoculate themselves against the most dreaded disease of the era with tobacco.

The momentum built. So many people seemed to be indulging in the leaf, and so often, and for so many reasons, that pipes, and occasionally cigars, could almost be confused for items of apparel, or at least indispensable accessories. So many Englishmen smoked at taverns and other places serving food that the leaf could be mistaken for a staple of diet. So many wreaths of smoke hung over so many heads, moving along with them, each like a visor or tiny roof, that the leaf could have passed for shelter.

"Children were permitted to smoke it, too," writes Peter Ackroyd in *London: The Biography*, "and 'in schools substituted tobacco for breakfast, and were initiated into the trick of expelling the smoke through their nostrils by their masters.' One diarist in 1702 recalled an evening with his brother at Garraway's Coffee House where he was 'surprised to see his sickly child of three years old fill its pipe of tobacco, after that a second and third pipe without the least concern.'"

Eventually, it would be said, a tobacco warehouse in one of London's outlying districts would be the largest enclosed space in the world except for some of the pyramids of Egypt.

Testimonials poured in—enough, perhaps, to fill that warehouse. The habitués of the Mermaid had returned to their inkwells, dipped in their pens and ran them across paper with all the enthusiasm they could muster. Ben Jonson had a character in one of his plays call tobacco "the most sovereign and precious weed, that ever the Earth tendered to the use of man." Spenser, in lines from *The Fairie Queene* that Raleigh had prompted, referred to tobacco as "divine." Christopher Marlowe, pedophile and poet, also chimed in: "All they that love not boys and tobacco are fools." And Samuel Rowlands, a contemporary who seems to have done most of his smoking and drinking somewhere other than the Mermaid, wrote an entire volume of light verse in honor of tobacco. A brief sample:

But he's a frugal man indeed,
That with a leaf can dine,
And needs no napkin for his hands,
His fingers' ends to wipe,
But keeps his kitchen in a box,
And roast meat in a pipe.

Some years later, luminaries such as Robert Herrick and John Milton would also rhapsodize on the weed's behalf, and years after that—in fact, centuries later—another famous wordsmith from the British Isles would make his predecessors seem restrained, even indecisive, claiming that

> with the introduction of tobacco England woke from a long sleep. Suddenly a new zest had been given to life. The glory of existence became a thing to speak of. Men who had hitherto concerned themselves with the narrow things of home put a pipe into their mouths and became philosophers. Poets and dramatists smoked until all ignoble ideas were driven from them, and into their place rushed such high thoughts as the world had not known before. Petty jealousies no longer had hold of statesmen, who smoked, and agreed to work together for the public weal. Soldiers

and sailors felt when engaged with a foreign foe that they were fighting for their pipes. The whole country was stirred by the ambition to live up to tobacco. Everyone, in short, now had a lofty ideal before him.

The author of the preceding is Sir James M. Barrie, and the account an even more fanciful one than his better-known exercise in imagination, *Peter Pan.* In fact, Barrie even saw fit to propose that England be renamed in honor of the man who so selflessly dedicated himself to tobacco's advancement. Raleighland, he would have his nation henceforth called! Was he kidding? More than likely, but it is worth remarking on the length to which he went in the process, and the zeal that must have motivated him.

Within a relatively short time after its arrival in England, smoking became the most popular leisure activity in the nation. One searches for an analogy, but in vain. Pipes and cigars cannot be compared to new toys, new games, new dances, or any other such diversion that, every so often, becomes the rage among a people. For these do not endure, do not continue to excite the interest or inspire the overheated prose. Tobacco won a following as speedily as does a fad but then remained a way of life. Never before had anything like that happened. The upper classes, used to novelty of one kind or another, eagerly succumbed to the weed; the lower classes, used to tedium, were frantic for novelty. Both groups found what they were looking for in Tobac-kee, and the king who succeeded Elizabeth to the throne would be furious.

IT WAS THE SAME IN OTHER COUNTRIES. Explorers conquered the New World, then returned home with tobacco, which conquered the Old. The leaf did not spread from one European nation to the next—not at the start, at least. It spread from America to one country, from America to another, from America to the next. The native tribes produced; the white man exported and accustomed himself to the new habit and consumed.

In all cases there was initial opposition, followed by centuries of acceptance, accompanied by outbreaks of further opposition. The nobility and those in their orbits were usually the first to light up; the peasants copied them and seemed to develop the stronger attachment, the barrenness of their lives increasing their susceptibility. People began smoking in a few carefully selected places. Soon they were doing it everywhere and at all times, and the more they smoked, the more avidly they found reasons to keep on with it, finally ascribing to tobacco the combined properties of Christ's burial shroud, penicillin, and a souvenir from the gift shop at Graceland.

It was the Spanish, oddly enough, who warmed most slowly to smoking. More than twenty years after Columbus returned from his first voyage, they were importing little of the weed, and much of what they did bring to their shores they utilized ornamentally, as plants to brighten the household rather than to ignite and inhale for pleasure or respite from illness. Perhaps the example of Rodrigo de Xerez still hovered, and they were afraid to risk similar misfortune at the hands of the church. Perhaps Columbus's sad last days had made his cargo, or his tales of the weed in America, seem inglorious.

In time, though, tobacco caught on in Spain, and among the reasons was a pamphlet called *Joyful News of our Newe Found World*, by Nicolas Monardes, a doctor in Seville, which "has never, even in the golden age of cigarette advertising, been surpassed in enthusiasm or excess." Because he was a physician, Monardes may have been an even more influential advocate for the leaf in Spain than Raleigh was in England. Referring to the American natives, whose testimony he believed without qualification, Monardes wrote, "They say [tobacco] is very good to drive forth and consume the superfluous moisture in the head. Besides, when taken in this way, it makes it possible to endure hunger and thirst for some time." He went on to claim that the leaf cured everything from kidney stones to tapeworms, from tiger bites to dandruff, from shortness of breath to diseases of the internal organs.

Jean Nicot, sent from France to Portugal in 1559 to arrange a royal marriage, added to the list. He was introduced to tobacco in Lisbon, where it had been given so spirited a reception that a ballet was composed in its honor. The setting was the West Indian island of Tobago, supposedly the homeland of the plant, and the performance began with the natives chanting their thanks to the gods for having allowed them to know the blessings that tobacco could bring. They were led by priests, who paraded before their subjects, exhaling deeply and often, turning the air thick and gray.

The second act featured the manufacture of various tobacco products, almost, or so it seems from an account of the time, in the style of a modern documentary. The audiences were apparently fascinated by this behind-the-scenes look at their favorite tokens of dissipation being created.

A dance that might have been a full-scale production number highlighted act three, with the participants dressed in a variety of costumes to represent a variety of nations, all of which enjoyed tobacco. This was the principle of the calumet writ large. Instead of merely calming the contentious members of a tribal council, the weed was bringing peace to the bellicose races of the entire world, a feat well beyond the skills, or even the wildest dreams, of statesmen who did not partake of the leaf.

Whether he saw the ballet in Lisbon or not, Nicot heard many a tale about both the taste and aftereffects of tobacco. The latter, he was assured, were salubrious. Initially, though, he was skeptical. He "determined to put the herb to the test. When he chanced upon a Lisbon man with a tumour he treated him with an ointment made from tobacco leaves and effected a complete cure." Making further experiments, and coming up with the same results, he wrote enthusiastically to the cardinal of Lorraine, who was also known by the title of Grand Prior:

> I have just got hold of a frightfully interesting Indian herb, which heals boils and running sores, which up to now seemed incurable. As soon as I have gathered some seed, I'll send it to your gardener at Marmoustier, together with some parts of the plant. I shall also add some instructions about how to grow it, just as I did when I sent you the orange-trees.

On his next trip to Paris, Nicot presented some leaves to his queen, the imperious Catherine de Medici. Neither boils nor sores were a problem for her, but headaches were, perhaps even migraines. She was a frequent sufferer and had gone so far in her search for relief, it is said, as to have ingested a powder made from ground mummies.

Thus, when Nicot told his monarch that tobacco was a better bet for what ailed her, she agreed to try it. That it worked, or that she persuaded herself it did, or that she enjoyed the stuff so much she did not even care about its efficacy, may be inferred from the terms by which the French came to speak of tobacco. In fairly rapid order, there were four of them: first, it was known as *Herbe du Grand Prior*, then *Herbe de la Reine*, then *Herbe Medicee*, and finally, as a tribute to France's answer to Sir Walter Raleigh, *Nicotiana*, which was the name that stuck.

It is no small irony that in later years, when a toxic alkaloid was found in tobacco—colorless and oily and repugnant to all the senses that encountered it, a substance that became synonymous with everything that was harmful and addictive about smoking, a substance that would one day be of value to society as an insecticide—it would be named after a long-deceased French politician who believed that the weed, far from being pernicious, was a bold new medicine.

Among other classes of French society, people for whom there was little in a name but much in tobacco to lighten the heart and enliven a dreary day, a tale soon made the rounds. It might even have been believed in one quarter or another. It seemed that a soldier known as La Ramée, though down on his luck, was still willing to share his last few coins with some other beggars, one of whom turned out to be Saint Peter in disguise. Delighted with his display of charity, Saint Peter pulled La Ramée aside and told him that he would grant him a wish, anything he wanted. He expected the man to ask for eternal life or great wealth or some beautiful, willing women—the old standbys.

But La Ramée surprised him. He needed only a moment to decide that the sum total of his desires was a square meal, which in some translations from the French is rendered as "tobacco and the food he saw in the inn."

So potent was the leaf, so strong was the hold it took of the imagination, so exalted was the status it had assumed in the Gallic society that it seemed only reasonable that La Ramée would wish for it and nothing else.

In another European land, and probably some years later, the plant was greeted in verse, a short poem about

a certain Count Herman
A highly respectable man as a German
Who smoked like a chimney,
And drank like a Merman.

The verse was a trifle, meant to amuse, but it hinted at a relationship between tobacco and alcohol, and people had begun to wonder. They would wonder all the more in the United States a few centuries later. Was there, in fact, a connection of some sort? Did smoking lead to a thirst for booze, or was it perhaps booze that led to a taste for the leaf? Might it be both? Neither? If one did not actually cause the other, did it at least provide some kind of encouragement or reinforcement for the other? If so, precisely how did that work? "Certainly the German's love of beer and tobacco is unrivalled elsewhere," writes F. W. Fairholt, a student of tobacco's past, and he is not disapproving:

> To drink a pint of beer at a draught and several quarts at a sitting; and to fill pipes as continually as they are burnt out, for the best part of a day, is no uncommon thing in Bavaria. It would astonish the weak minds of tee-totallers and tobacco-haters could they take a seat for a day in any *Lustgarten* in this most philosophic nation.

In truth, the German love of beer and tobacco *was* rivaled elsewhere, at a number of elsewheres. Raleigh somehow knew that those at the Mermaid, quaffing their ale and swilling their wine, were likely allies in his quest to promote the weed. He did not try to push it at a convent or a trading house.

And in other countries, as well, there were more and more individuals finding the tastes of tobacco and brew—or, in some cases wine—to be quite complementary.

In the Netherlands, for instance, the person who eschewed both pipe and grape was looked at as an outcast, vilified in a poem that many found insulting ...

Die geen toeback op wijn mag lijden,
Ick whens hy dag en nacht mag op een esel rijden.

... but that, admittedly, loses much of its bite in the translation:

Whoever does not like the weed or wine,
May he go donkey-riding day and night.

Few, though, were the Dutchmen consigned to such a fate. Smoking and drinking were seen as partners in the causation of venery and idleness, which to some of the citizenry was reason for hand-wringing but to a large number of others seemed a mutual endorsement. Regardless, writes the historian Simon Schama, "The smell of the Dutch Republic was the smell of tobacco." Schama continues with a description that is also accurate, at least in general terms, of an earlier age:

> In the middle of the eighteenth century the French traveler Grosley counted three hundred smokers in a single modest inn at Rotterdam and complained that the Dutch were so indifferent to the poisoning of confined spaces that the fumes from traveling barges drove foxes from their lairs as they passed.

But the same fumes drove merchants to ecstasy. They were selling tobacco so quickly and profitably that at least a few of them had a hard time keeping up with demand and counting their money. Even so, they tried to sell more, hoping to attract the casual or beginning smoker by marinating the leaf and mixing it with spices, in much the same way that today's upscale coffee bars add all sorts of unexpected flavors to the bean. Among the spices most commonly used were thyme, dill, lavender, and nutmeg. Other marinades included prunes, vinegar, and beer.

In Brussels, King Frederick William I, the father of Frederick the Great, became so enamored of tobacco that he was seldom seen without it. His subjects began to call him "The Smoking King." And his weekly assemblage of unofficial advisers, pipe smokers all, was soon known as the *tabaks-collegium*, or smoking parliament.

The Japanese were no less welcoming. "Samurai knights formed smoking clubs," it has been said, "and commissioned elegant paraphernalia in a manner reminiscent of the Elizabethan 'reeking gallants.' They favoured ornate silver tobacco pipes which they strapped to their backs or tucked into their kimonos beside their legendary swords."

In China, tobacco was thought to be a cure for malaria. Writes a physician in the service of the emperor at the time about Yunan Province, "When our forces entered this malaria infested region, almost everyone was infected by this disease with the exception of a single battalion. To the question why they had kept well, these men replied that they all indulged in tobacco. For this reason it was diffused into all parts of the country. Everyone in the South West, old and young with exception, is at present … smoking by day and night." In addition, tobacco "cures troubles due to cold and moisture, removes congestion of the thorax, loosens the phlegm of the diaphragm, and also increases the activity of the circulation," all of this despite the fact that it "has an irritating flavor and warm effect and contains poison."

Of Persia, it was believed that the citizenry would rather smoke than eat, and the exaggeration seems to have been slight. Here, as in other nations, the royals were the ones advertising the weed but with a single, important exception: Shah Abbas the Great, an admirable man who is credited with modernizing the Persian army, building roads and bridges, and stimulating trade with as many other nations as possible.

What he did not wish to stimulate was the desire for smoking. He was, in fact, appalled by the extent of it among his people, especially his subjects at court. He swore that he would stop it. But the shah was not a despot, or even comfortable in confrontation. Rather than ordering his courtiers and attendants to put away their pipes, and thereby encourage them to light up behind his back and in the process lose respect for the sovereign word, he decided on a trick:

> He had horse manure dried and substituted for tobacco in the vessels from which they filled their pipes; he explained that this was a costly product presented to him by the vizier of Hamadan. They smoked it and praised it to the skies; "it smells like a thousand flowers," vowed one guest. "Cursed be that drug," cried Shah Abbas, "that cannot be distinguished from the dung of horses.

There were more people blessing tobacco than cursing it, however, not only in Persia and the Far East and most of Europe, but even, after a time, in Africa, where Ugandan priests took up the pipe for the same reasons, and with the same devotion, as had the Mayas. As a result, it soon became necessary to construct myths about the weed's discovery. A myth, after all, is not merely a tale meant to entertain, not just a fiction employed because the truth us unknown or insufficiently compelling. It is, instead, a monument made of words, a means of attesting to the rarefied position of its subject. It does not deal in trivia or irrelevance. Rather, it explains the life of a deity or a basic quality of human nature; it reveals the founding of a country or a school of thought or the creation of a natural wonder. Tobacco, of course, was regarded by many as the latter.

Actually, several nations developed myths about tobacco, although no one can quite describe how they came about. The stories always seemed to be there, just as features in the landscape always seemed to be there. Each myth was different from the others, but most had a few plot elements in common. Usually a man, possibly a holy man, chanced upon a god or an animal with the gift of speech. The two talked for a while, then an argument broke out, often followed by a violent action of some sort that inadvertently led to tobacco's taking root, in both the soil and the society. One of these myths, perhaps the prototype, is charming enough to deserve a full retelling:

> The Prophet was taking a stroll in the country when he saw a serpent, stiff with cold, lying on the ground. He compassionately took it up and warmed it in his bosom. When the serpent had recovered, it said:
>
> "Divine Prophet, listen. I am now going to bite thee."
>
> "Why, pray?" inquired Mahomet.
>
> "Because thy race persecutest mine and tries to stamp it out."
>
> "But does not thy race, too, make perpetual war against mine?" was the Prophet's rejoinder. "How canst thou, besides, be so ungrateful and so soon forget that I have saved they life?"
>
> "There is no such thing as gratitude upon this earth," replied the serpent, "and if I were now to spare thee, either thou or another of thy race would kill me. By Allah, I shall bite thee!"
>
> "If you hast sworn by Allah, I will not cause thee to break thy vow," said the Prophet, holding his hand to the serpent's mouth.
>
> The serpent bit him, but he sucked the wound with his lips and spat the venom on the ground. And on that very spot there sprung up a plant which combines within itself the venom of the serpent and the compassion of the Prophet. Men call this plant by the name of tobacco.

Smoking was so unconventional a custom, yet it became so commonplace, so quickly. Men, and a surprising number of women, lit their pipes or cigars and relived the pleasures of the nipple as they pulled the smoke as deeply into their bodies as they could. They felt it make its way through their internal routes and byways, felt themselves stirring, tingling in places that had never tingled before. Then they blew out the smoke and stared at it as it floated away from them, perhaps musing on its destination and purpose, perhaps not. The world was still a large and terrifying place, and much that happened in it remained inexplicable. Tobacco was for some a small mastering of the environment, a few precious minutes of control, an assertion by morals that the cosmos was their home and they would not be made to feel alien within it.

To others, it was a habit so filthy—so utterly despicable, if not even dangerous—that something had to be done about it, drastically and finally and soon.

TWO

The Enemies of Tobacco

THE ROOTS OF EVERY anti-smoking sentiment ever uttered and every anti-smoking document ever published and every anti-smoking movement that ever existed may be found in the reactions of white men to their first sight of tobacco. They may be found in the confusion of de Torres and de Xerez and in the downturned lips of Cartier and in Nicholas Perrot's amazement as his head was engulfed in fumes and his ears filled with the sounds of supplication. They may be found in laughter and curses, in whispered questions and shouted disdain, and in the sometimes comical attempts of those who witnessed tobacco in action to explain the procedures to those who had never seen or heard of them before.

Some of the witnesses became smokers themselves. Some became the opposition. The former always outnumbered the latter, at least until the twentieth century, but the two groups marched together in lockstep, force and counter-force, partners through history in the same way that saints are partners with sinners, that hedonists are in league with ascetics. One group needed the other. One group defined itself, at least in part, by its struggles with the other. The reformers were motivated, not intimidated, by tobacco's popularity. It would not have been worth their while to take up arms against a custom that mattered to only a few. Smokers, however, were determined not to be put off by scolds, not when their habit was as beneficial as it was blissful.

The roots of opposition to smoking, it may be said in summation, are to be found in the clash between the practices of one culture and the perceptions of another.

That the Europeans and the native Americans had nothing in common was obvious to both from the beginning. They spoke different languages, wore different clothes, worshiped different gods. They ate different foods and lived in different kinds of abodes. They moved through life at different speeds, expecting different outcomes, and taking different approaches to bring them about. The Europeans had one set of physical features, the natives another; the European temperament was a puzzle to the American tribes, and the same was true in reverse. The two cultures bore every sign of having been separated by a vast body of water for virtually all their existence.

But most important as far as attitudes toward tobacco were concerned, the Europeans saw themselves as the more advanced civilization. It was they, after all, who had reached the New World and discovered its inhabitants, not the indigenous peoples who had sailed east and explored England, France, and Spain. In science and art, philosophy and invention, government and economics, the Europeans accomplished far more than did the American natives. As a consequence, they produced societies that were more fruitful and efficient and likely to endure, and the stores of knowledge they passed along from one generation to another, in ever increasing amounts, were an incomparable legacy, one that ensured the future as it paid homage to the past. Apart from instinctive objections, then, a certain number of Europeans looked down on smoking as the pastime of less sophisticated human beings.

As such, it raised a variety of questions, and each was a rendering of judgment as much as a request for information. Why would a person deliberately hold fire so close to his body? Why would he deliberately bring the smoke inside him and take his time about releasing it? Why would he allow himself to smell like tobacco, which to many residents of the Old World suggested the more unpleasant odors of urban living? Why would he willingly subject himself to the taste, which was equally repugnant, like food that has gone bad or a plant not fit for human consumption? Why would he assign to such a substance spiritual and medicinal powers, when it was so obvious that tobacco possessed neither? The entire picture of a man or woman smoking a cigar or pipe was an enigma to the European at first. It seemed not only a pointless thing to do but an unnecessary risk.

Attitudes like this brought at least a few white men to the brink of concluding that tobacco was harmful to one's health. But they did not go that far, not yet, not until early in the seventeenth century. They did not attribute any specific diseases to the weed, nor could they see exactly how it might cause the body to fail in its operations. But there were those who began to wonder in general terms whether tobacco might lead to a lack of physical vigor. Heat, after all, was often debilitating to a person, and if that heat were taken directly into the body, would it not turn a person lethargic?

At the very least, some people believed that the claims they had heard about the weed's curative properties were overrated. After all, no other healing agent of the time was as malodorous with decay, nor was any administered in so patently offensive a manner. Yet the New World tribes swore that tobacco was the most magical medicine of all, a medicine to cure every disease, to eliminate every curse.

In the early days, of course, the anti-smoking movements were not movements at all. They were a few voices here, a few there, a few somewhere else, most not even aware that the others had been raised. It would be a long time before they formed themselves into a chorus, and even longer until a majority of other men and women paid attention to their tune.

ONE OF THE first Europeans to complain about tobacco in writing was Girolami Benzoni, an Italian who visited the New World in 1541 and took almost immediate offense:

> See what a wicked and pestiferous poison from the devil this must be! It happened several times to me that, going through the provinces of Guatemala and Nicaragua, I have entered the house of an Indian who hath taken this herb, which in the Mexican language is called *tabaco*, and immediately perceiving the sharp, fetid smell, I was obliged to go away and seek some other place.

Benzoni was not just relating a personal experience; he was warning others who might be tempted to try the weed, meaning to scare them off, and so he exaggerated, going on to claim that he had seen some Indians take in so much tobacco "that they fall down as if they were dead and remain the greater part of the day and night unconscious." He acknowledged that the natives "feel a pleasure" in this stupefaction but was mystified as to why. He felt no such rapture himself.

There are not many documents like Benzoni's from the sixteenth century. It was a time when few men set down their impressions about anything, much less tobacco, and even fewer of the impressions survived the years. It seems safe to assume, however, that Benzoni was not alone in his criticisms, and even safer to say that, once tobacco showed up on the other side of the Atlantic, cargo of the conquerors, even more people objected. Surely poems were written to denounce the weed as well as to welcome it, and some ladies and gentlemen must have walked out of the ballet in Lisbon as the others stood and cheered. Surely there were those not just in the Netherlands but in all nations who would rather have ridden a donkey than smoked a pipe, regardless of what their countrymen thought. And surely one or two members of the tabakscollegium found it hard to breathe at their sessions with the king, and one or two of the Mermaid Tavern habitués chose to keep drinking or retire to the writing desk instead of applying flame to pipe and nodding toward Raleigh.

Even Ben Jonson seems to have admitted some doubt. Although not directly contradicting his previous statement that the leaf was "sovereign and precious," Jonson put other words into the mouth of Captain Bobadill, a character in his play *Every Man in His Humour*. The captain is a smoker and seems to be proud of it. Still, he refers to "roguish tobacco ... good for nothing but to choke a man, fill him full of smoke and embers." And Samuel Johnson, more than a hundred years later, would speak for many who had gone before him. "It is a shocking thing," he said, "blowing smoke out of our mouths and into other peoples' mouths, eyes, and noses, and having the same thing done to us."

Also shocking to some was the cost of the weed. Edmund Gardiner's 1610 work, *The Triall of Tobacco*, laments the fact that the "patrimony of many noble young gentlemen have been quite exhausted, and have vanished cleane away with this smoky vapour, and have most shamefully and beastly flyen out at the master's nose." Complaints like this were not unusual. Smoking was not just a new activity for human beings but an entirely new category of expense, resented for that reason alone by those who either could not afford it or could imagine better uses for their income. Gardiner had to admit, though, that the leaf was "a fantasticall attracter." He was not surprised that so many young men had a hard time resisting.

Still others worried not about danger to taste or morals or financial well-being but to infrastructure. Seventeenth-century Europe was built largely of timber: houses and barns, shops and sheds, and workplaces of all sorts were vulnerable to a single, careless smoker. So were a number of items found commonly in such places. What if a person dropped his cigar on a stack of papers or a bale of hay? What if he placed his pipe too close to another's vest or petticoat? What if he fell asleep with his smoke still smoking and it dropped into the straw of his mattress? This kind of thing did not happen often, but there was always the possibility. Even before much of the city of London went up in flames in 1666 "it was decreed that each new sheriff and alderman, within a month of taking up office, 'shall cause 12 new buckets to be made of leather for the quenching of fire.'"

Churches, the grander ones at least, were made of stone. Nonetheless, it was the clergy who led the opposition to smoking in the Old World, finding it a distraction from holy objectives and, more to the point, such a source of earthly delight that it challenged one of the church's fundamental teachings: that a person must sacrifice his joys in this life to be rewarded in the next. By doing so, of course, tobacco was a threat to clerical dominion, an instrument of heresy, to be avoided by true believers as they would avoid any other temptation to stray.

In 1583 or thereabouts, Pope Sixtus ordered his divines not to make use of the leaf in any way during services. "Under pain of mortal sin," read an edict of his, "no priest, before celebrating and administering the Communion, should take tobacco in smoke or powder not even for medicinal reasons." In 1624, one of his successors, Urban VIII, complained that "the use of the herb commonly called tobacco has gained so strong a hold on persons of both sexes, yea, even priests and clerics," that he banned it from all Catholic churches at all times. He was especially concerned, he made known, with soiled altar linens and the tendency of noxious fumes to divert attention from worship by lingering in the air long after a pipe or cigar had been extinguished. One cannot simultaneously wrinkle up his nose in disgust and concentrate on adoration of the Almighty.

Some years later, a papal bull demanded that Catholic men of the cloth not use tobacco in any form, at any time, or in any venue of Seville. And a few decades after that, Innocent XII went even further, excommunicating all those who "take snuff or tobacco at St. Peter's in Rome."

When, in the following century, Benedict XIII ruled that excommunication was too harsh a penalty for that offense, he also stated that "the revocation was not to be taken as license." He, too, wanted tobacco to be kept separate from the tabernacle, filth to be as far removed as possible from supreme cleanliness.

A man named Benedetto Stella, and perhaps he alone, disagreed with such clerical decrees. Writing in 1669, Stella made a connection between the weed and celibacy:

> I say ... that the use of tobacco, taken moderately, not only is useful, but even necessary for the priests, monks, friars and other religious [*sic*] who must and desire to lead a chaste life, and repress those sensual urges that sometimes assail them. The natural cause of lust is heat and humidity. When this is dried out through the use of tobacco, these libidinous surges are not felt so powerfully.

Had the Mayas of earlier times known all of this—in fact, had they known *any* of it—they would have dismissed the notion of the white man's superiority without a second thought. How, they would have asked, could a human being expect the gods' favors when he refused their most precious gift? How were prayers to reach the heavens without smoke as a means of transport? How could a man hear the celestial responses without tobacco to attune his consciousness?

Among the monarchs of the time there was also a growing resistance to the leaf. Queen Elizabeth did not share it, but many of her fellow throne sitters did, and for the same reason as the clerics: It was a threat to their hegemony, a sign of independent decision making on the parts of men and women whose subservience was required for the smooth operation of the kingdom. Shah Abbas, then, was not alone in his attempts to discourage the use of tobacco among his people. He does, however, seem to have stood by himself in the civility of his methods.

In Russia, Tsar Michael Feodorovich instituted a sliding scale of torture. A first-time smoker was whipped with leather thongs until bloody and repentant; after a second conviction his nose was slit, and if he was caught applying flame to tobacco again, he would have his head removed in a ceremony to which one and all were invited. In 1641, the more humane alternative of exile was added to the list of punitive options, although it does not seem to have been utilized very often, perhaps being too much trouble and expense. A few years later, the Russians imposed severe penalties with more frequency after a series of fires broke out in Moscow.

The seventeenth century was not a good time to be a smoker in Turkey, either, even though the land would one day rank among the world's leading growers and exporters of leaf, its name synonymous with the smoothest blends and richest tastes. A traveler whose name is not known to history told of seeing "an unfortunate Turk conducted about the streets of Constantinople in 1610, mounted backward on an ass with a tobacco-pipe driven through the cartilage of his nose, for the crime of smoking."

Things would get worse. In 1633, Murad IV, also known as "Murad the Cruel," made tobacco use a capital offense in Turkey. His method, we are told, was entrapment. He would wander through Constantinople in disguises of various sorts, ask those whom he encountered for some leaf, and then order those who complied to be beheaded. According to one account, 25,000 Turks lost their lives for this reason in less than a decade and a half. It should also be pointed out, though, that Murad the Cruel executed coffee drinkers, wine drinkers, opium takers, and all of his brothers save one. He is hardly typical of the sultanate, and tobacco was not the only substance repugnant to him. Still, as late as the nineteenth century, the Turks were cutting off the noses of smokers rather than ramming pipes through them—not exactly a sign that the weed was looked at benignly or that legal codes had become more humane.

Rulers in China and Japan took the lives of smokers, or threatened to, with the Japanese also seizing the smokers' property and handing it over to the military. In Transylvania, confiscation was the penalty for those who grew tobacco. People who actually smoked the stuff were fined as much as three hundred florins. Jahangír, the Mogul emperor of Hindustan, decreed that smokers have their lips slit so that a pipe would never again rest comfortably between them.

Even Shah Abbas's own grandson behaved barbarically, once commanding a minion to pour molten lead down the throats of two people who smoked in public. Not for him a hearty little joke like horse dung in the leaves.

Other countries passing laws to eliminate or curtail the spread of tobacco in the first half of the seventeenth century included Sicily, Denmark, Germany, Austria, Hungary, the Papal States, the Electorate of Cologne, Wurttemberg, and Switzerland, where smoking was perceived in statute as only a little more commendable than adultery, and the two were treated in a similar manner. Punishments in these lands ran the gamut from death to fines and from disfigurement to the confiscation of material goods, although enforcement was often spotty.

Still, said F. W. Fairholt, it was an overreaction, such an overreaction, so much legislative force directed against so unassuming a foe:

> Was ever the destruction of body and spirit threatened so unjustly? Mutilation for taking a pinch! Loss of life for lighting a pipe! Exclusion from heaven for perhaps harmlessly reviving attention to a wearisome sermon in chapel or church! Merciful heavens! What comminations these to emanate from Christian kings and Christ's successors!

Although he was by no means the harshest monarch of his time, the Christian king of England might have been the most fanatical in his hatred of tobacco. He surely devoted more time to it than did other leaders: contemplating, fulminating, recording his sentiments on paper, and forcing his will on Parliament. He had only to wait for his predecessor to pass away before his personal obsession became at once the national policy and the national bane.

ELIZABETH I COULD be a patient woman when she put her mind to it, and this was never more true than when she prepared herself to die. She took her sweet time about it, did the beloved sovereign, allowing life to slip from her gradually, in almost imperceptible degrees. Her subjects watched with leaden hearts, bracing themselves as best they could. They remembered all that was noble in Elizabeth's tenure and considered themselves the better for it, both as a nation and as individuals. They hoped she was on her way to a better place, and when she finally departed, on March 24, 1603, one could hardly walk past a home or shop in London without hearing someone sobbing within. Toasts were drunk and eulogies offered by people of every rank, at occasions of every sort or at no occasion at all.

It was more than a month, though, before Elizabeth was buried in Westminster Abbey, and on that day thousands of lords and ladies marched slowly behind the funeral bier, their heads down, cheeks damp, not uttering a word. Bringing up the rear of the procession, Giles Milton writes, "was an aged, limping, but still handsome courtier. Sir Walter Raleigh was leading the Gentlemen Pensioners in their mourning, their gilded halberds pointing at the ground in token of their grief. All knew that this was the end of an era: 'her hearse (as it was borne) seemed to be an island swimming in water, for round it there rained showers of tears.'"

It was not just that the nobles missed the queen. They were also dubious about her successor, reluctant to acknowledge him, even though he was Elizabeth's own choice, not to mention her first cousin, twice removed, the son of Mary Queen of Scots and her second husband. He had previously ruled Scotland as James VI. On moving south and assuming his new throne, he became James I, the first Stuart king of England.

He was a man about whom few have ever held a neutral opinion. Some admired him for his steadfastness, others for his learning. His book *Airs of Scottis Poesie* was well regarded by poets and scholars alike, and, in an entirely different vein, he produced two volumes about the generally hostile relationship between the crown and the papacy: *The True Law of Free Monarchies* and *Basilikon Doron*, which means "the royal gift." James's Anglican faith, which inspired the latter tomes, was a beacon to him, and it may be that he enjoyed nothing more at the end of a hard day's ruling than to discuss his religion in a quiet setting with friends of like mind. It was he who sponsored the translation of the Bible known as the King James version, a treasure for both literary and spiritual reasons, and he took great pride in the accomplishment.

He was also proud of keeping his nation at peace for the twenty-two years of his stewardship. There were a few outbreaks of fighting with Spain but nothing that could be called warfare. He even signed a treaty with the Spaniards in 1604, and considering the constantly frayed relations between the two countries, this was a genuine feat of diplomacy. Among his other accomplishments were introducing golf to England and providing gainful employment for the nation's most brilliant architect, Inigo Jones, whose Banqueting House at Whitehall Palace, erected but a few years before James died, was not just one of the great buildings of the time but one of the great artistic accomplishments of any kind.

However, the king did not wear his scholarship lightly. He was, rather, "a man of ponderous erudition" who "lectured Englishmen on every topic but remained blind to English traditions and sensibilities." A chronic invalid, and perhaps a hypochondriac, he was possessed of legs too spindly to support his torso and a tongue the size of an adult eel. It was this, thought his detractors, that accounted for all the king's drooling, as the creature kept slipping out of his mouth and running down his chin, despite his best efforts to confine it. The tongue also seems to have been responsible for the difficulty James had in consuming food: "It was said to be possible to identify every meal he had eaten for seven years by studying the scraps of dried food stuck to his clothes."

James was accused of playing favorites more than did most monarchs and to have spent lavishly on those who had earned his esteem. "He also spent plenty of money on himself," says Jane Murray, "foolish though it may seem for him to have bothered with velvets and jewels when he never bothered to wash his hands." Even less kindly, there were those who called attention to "his big head ... codfish eyes ... his want of dignity, his drunkenness ... and his rank cowardice."

He was not cowardly in assailing tobacco, however. It was the one war he *did* declare, and he pursued it vigorously, this British potentate

In quilted doublet
And great trunk breeches,
Who held in abhorrence
Tobacco and witches.

It is not clear whence the abhorrence for witches came, and why it prompted him to execute as many as four hundred in a single year of his Scottish reign. Nor is the source of his disdain for tobacco known, although he might have perceived the weed as a product of witches and the satanic forces they worshiped. Certainly, James had a sincere dislike for it: Tobacco offended him for the same reasons that it offended many others. But he might also have been playing politics, at least at the outset. He might have chosen smoking as an issue on which to make a stand to his new subjects, to show them he was his own man and would be a leader of firm resolve, that he was not to be taken lightly as a newcomer or dismissed altogether as nothing more than the former queen's kin, or even puppet. In the process, his opposition to tobacco might have assumed a life of its own, becoming more than he initially intended it to be. He might, in other words, have found himself typecast in a role of his own creation, although a role true to his nature.

Regardless of how it happened, James was obsessed by the weed, and shortly after taking over for Elizabeth he produced the most famous anti-smoking tract of his era, the *Counterblaste to Tobacco.* It was not the first such publication. Among writers of English, that distinction probably goes to the pseudonymous Philaretes, whose 1602 opus, *Work for Chimney-sweepers,* claimed that the leaf would make the brain so sooty that an entire army of the title characters would be required to cleanse it.

But history has forgotten about Philaretes, whoever he was. The *Counterblaste,* however, is a landmark in polemical literature. In fact, not until 1964, when Luther Terry, the surgeon general of the United States, released his report on the weed and its various consequences, was there anything to equal James's opus in impact. Although he denied authorship for more than a decade, there was never any doubt that the *Counterblaste* was the work of his very own pen, dipped in his very own brand of vitriol.

It is, by any standard, a curious piece of work. In places, the king's arguments are well expressed, perceptive, and logical in their movement from one point to the other. In addition, they covered a great number of points, James leaving no stone unturned if that stone could be hurled at tobacco.

But in other places, it is impossible to make out the author of *The True Law of Free Monarchies* or *Airs of Scottis Poesie.* So intemperate is the language, so rancorous and vituperative, that the *Counterblaste* occasionally reads like a spur-of-the-moment screed by a man who has been provoked beyond endurance.

He begins with an attack on those who invented smoking, putting the cultural conflict in its most unflattering terms. "What honor or policie can move us to imitate the barbarous and beastly manners of the wilde, godlesse and slavish Indian," he asks, "especially in so vile and stinking a custome?" He goes on to

> say without blushing [why do we] abase ourselves so farre as to imitate these beastly Indians, slaves to the Spaniards, refuse to the world, and as yet aliens to the holy Covenant of God? Why do we not as well imitate them in walking naked as they doe? ... Yea, why do we not denie God and adore the Devill, as they doe?

Further along, James discusses the effects of tobacco on the smoker, prescient about what later research would find and, perhaps, revealing that he had taken a glance or two at Philaretes. He claims that smoking

> makes a kitchen of the inward parts of men, soiling and infecting them with an unctuous and oily kind of soot. ... Is it not a great vanity, that a

> man cannot heartily welcome his friend now, but straight they must be in hand with tobacco? ... [T]hat the sweetness of man's breath, being a good gift of God, should be willfully corrupted by this stinking smoke?

The man who inhales tobacco will also find, James cautions, that "his members shall become feeble, his spirits dull, and in the end, as a drowzie lazie bellygod, he shall evanish in a Lethargie." And the effects of the weed on the individual will mirror those on society as a whole, for to James's dismay he already saw in his subjects "a generall sluggishnesse, which makes us wallow in all sorts of idle delights."

The king even gets satirical at one point, trashing the notion of the weed as medicine. "O omnipotent power of *Tobacco*!" he sings, meaning precisely the opposite, incredulous that people could confuse misfortune's cause for its remedy.

In conclusion, James reaches a crescendo of animus, referring to smoking as

> a custome lothsome to the eye, hatefull to the nose, harmefull to the braine, dangerous to the lungs and in the blacke stinking fume thereof, nearest resembling the horrible Stigian smoke of the pit that is bottomlesse.

Smokers were as bewildered by the *Counterblaste* as they were upset. They asked themselves why the king was so angry. They asked why the anger was directed at so modest an amusement as smoking. Was James losing his royal mind so soon after becoming their ruler? They cursed him in private conversations and in broadsides that were passed cautiously from hand to hand in shops and taverns and on the street. They told jokes about him, all of them meanly intended, and when they lit their pipes and cigars and sucked the smoke into the private chambers of their being, they did so in defiance as much as satisfaction. Take *that*, James the First of England, each drag seemed to mean. If only the flames that ignited the tobacco would consume the wretched *Counterblaste*, reducing it to ashes, dismissing it from memory.

The citizens of the realm missed good Queen Bess more than ever, and at least a few of them sought consolation from the fact that James had begun his rule at the age of thirty-seven, relatively late in life for a man in those days. Perhaps he would not be around much longer to torment them. Perhaps he would begin to decline soon, tomorrow or the next day. Perhaps his successor would be a three-pipe-a-day man.

The king did not mind the furor. He even thought about ways to stir it further. Not content merely to have committed his bile to paper, James considered making tobacco illegal, discussing the possibility with his advisers. But they warned against it, reminding James as delicately as possible that he was already having a problem with his image. In his relatively brief time on the throne, he had alienated Parliament over the issue of Divine Right, his own Anglican church over the issue of lay participation, and the Catholic church over whether it was even entitled to exist within British borders. He could not, his advisers insisted, risk alienating smokers, as well, since that group contained members of the other three in addition to much of the rest of his new domain. Besides, they told him, a ban on tobacco would be almost impossible to enforce. Attempts to do so would not only fail but consume time and money at fearful rates. Reluctantly, James agreed.

But he would not be talked out of raising the tax on tobacco by perhaps the most that any tax on any single item has ever been raised at a single swoop. From two pence per pound it went to six shillings and ten pence, an increase of 4,000 percent! This applied both to imports, which made up the great majority of weed smoked in England, and to the domestic variety, which had been grown for only a few years, primarily in Gloucestershire and Worchestershire, and, not tasting nearly as good, was regarded by most as a poor man's consolation.

Furthermore, James ordered merchants who sold tobacco and the tradesmen who manufactured pipes to pay a special fee for a license, something they had never had to do before. This course of action, Jerome E. Brooks says, had the expected result, as it

> immediately intensified the activities of smugglers ... enraged the honest trader, and seriously reduced the income to the customs. And it had the usual effects of interference with a popular habit; both the "persons of mean and base condition" as well as "the better sort" fell more furiously than ever to smoking.

Eventually James lowered the tax, although not to previous levels. He also began to control his temper and moderate his language when it came to the weed—in public, at any rate. He wrote no more counterblasts and stopped referring as much to the one that had already been written, turning his lordly attention to other matters of import to the kingdom. When he acted dispassionately, he often acted wisely. Perhaps he knew he had gone too far.

But the damage had already been done. His violent outbursts against tobacco had made England's smokers his enemies, and they would neither forgive nor forget. James, in turn, frustrated that his more restrained self seemed not to improve his standing, seethed more than ever. This might explain, at least in part, his willingness to chop off the head of the nation's most popular advocate of the leaf.

IT SHOULD HAVE BEEN the happiest day of Sir Walter Raleigh's life, and in many ways it was. In 1588, he married Elizabeth Throckmorton, thirteen years younger than he, "no heiress, and no great beauty . . . a fair-haired, blue-eyed, nice girl, with a pleasant face and figure." He had courted her for some time, inviting her to "enjoy sweet embraces, such delights as will shorten tedious nights."

But one night had been too short, too delightful, and as a result of it Elizabeth was pregnant by Raleigh when they wed. Afraid of the reaction at court, he persuaded her to keep their union, and their child, a secret for as long as possible. She was quick to agree, for Elizabeth the bride was one of the ladies-in-waiting to Elizabeth the queen, and the regent was as protective of her attendants as a parent, if not more so. She shielded their virtue "with a fierceness that verged on paranoia. She treated every smirch on their purity as though she herself had been despoiled."

Elizabeth Throckmorton, of course, was as smirched as a woman could be in sixteenth-century Europe. That was bad enough. But when the queen learned about her condition, as she inevitably did, and then discovered that Walter Raleigh—of all people—was responsible, that both deed and deception were her old soulmate's doing, she was aghast. It was a blow from which their relationship never recovered. She thought he had betrayed her; he thought she had been too quick to judge, novice that she was in affairs of the heart. Or flesh. Such was the chasm that opened between her highness and Raleigh that he and his wife spent what amounted to their honeymoon in the Tower of London, imprisoned for several offenses, real and otherwise, all of which boiled down to the circumstances of their marriage and the queen's raging disfavor.

Its shaky start notwithstanding, the marriage was a good one. Raleigh doted on his young Elizabeth, spending as much time with her and their children as he could, and when his travels parted them, he wrote her poetry of a more passionate nature than he had ever composed for the queen. He was never sorry to have loved young Elizabeth, not for a moment, and he never thought that the illegitimacy of their firstborn was in any way a smirch on father, mother, or child.

But Elizabeth I still mattered to him. Somehow Raleigh had to repair the breach, win back her regard, not to mention his own standing at court. He was miserable in his role of outsider; nothing in his nature had prepared him for it. But what could he do? How could he persuade his monarch to forgive him? It was a question he asked himself countless times, often discussing it with his wife, the two of them sitting there in the Tower of London in the early days of their union, trying to come up with a plan for both their sakes.

They did not. The queen, however, did. Simply put, Elizabeth ordered Raleigh released from confinement so that he could deal with thieves who, she believed, had looted British trading vessels of spices and other valuable goods. Raleigh accepted the mission eagerly. He found the thieves, saw to their punishment, and restored Elizabeth's treasure to her. But she rewarded him only with his freedom, not a share of the spoils, as he probably expected. Raleigh and his wife left the Tower of London shortly before Christmas 1592, still disgraced and now beginning, for the first time, to worry about money.

The next chance for redemption would not come until a few years later. The queen summoned Raleigh again and told him he was to lead an expedition to the New World in search of El Dorado, the fabled city of gold that no one had ever seen but in which everyone seemed to believe. She was still distant from him, still bitter about his betrayal of her, but she told been told by others, and believed herself, that he was the only man for the job.

He agreed, but with an ulterior motive. Raleigh would not only try to win back the queen's favor with his voyage but would seek to fulfill a desire of his own, as visions of El Dorado had long since danced in his head, capturing both his imagination and his lust for riches. He might have believed, as did others at the time, that the walls of the city were "sheathed in slabs of gold." He might have believed that "the gold would reflect its light so brightly that [the entire city] glowed in the middle of the lake, appearing to be a second sun rising from the lake waters." He might have believed that even the houses of poor people in El Dorado were made of silver. With the charge from his queen, then, he could act on his beliefs and, in the process, regain both wealth and reputation.

Raleigh prepared diligently for his quest. He assembled the most capable crewmen and the most modern equipment available. He studied maps and documents, devising a strategy of exploration, and talked to others who had traveled to various parts of the Americas, comparing their impressions with those he had collected on his own journeys. He imagined what he would find, imagined how he would react, tried to eliminate all possibility of surprise, especially the kind that might leave him at a disadvantage. When he finally set out for the New World, it was with sails full and expectations fuller.

The mission was a flop. If El Dorado was not a fiction, it remained an undiscovered reality. Raleigh left England in February 1595 and returned in September with no riches, few mementos, and his name now more notorious than famous. In fact, he returned to face rumors that he had never even left, the story being spread by foes that he had spent the entire nine months of the alleged voyage "skulking in an obscure cove in Cornwall."

In truth, Raleigh had journeyed to the South American country of Guiana, which intrigued him in many ways, and he would eventually win a degree of literary renown by writing about it in *The Discovery of the Large, Rich, and Beautiful Empire of Guiana with a Relation of the Great and Golden City of Manoa*, the very title an attempt to justify his journey, and its costs, to his queen. But she was not impressed. She was even less impressed with Raleigh's pleas that the British establish a colony in Guiana—that they, in effect, build their own El Dorado. This she told him, and then sealed her lips. The ruler and her subject were now more estranged than ever.

By this time, the queen's health was beginning to fail, and thoughts of succession had entered the minds of many. Among them was Raleigh's principal antagonist at court, Lord Henry Howard, first earl of Northampton. If he had not made up the Cornwall cove canard himself, Howard certainly took pleasure in passing it along and swearing to its authenticity. In his opinion, Raleigh was one of the most duplicitous human beings in all of England, a man who "in pride exceedeth all men alive" and one who was easily "the greatest Lucifer that hath lived in our age."

Howard also believed that because of the Guiana fiasco, Raleigh now stood on the brink of ignominy. He decided to push. Assuming that James VI of Scotland would occupy the British throne next, although no public announcement to that effect had yet been made, he wrote to the future king, introducing himself as a friend and Raleigh as a man who could not be trusted under any circumstances and should not be among James's advisers. He continued:

> Let me, therefore, presume thus far upon your Majesty's favour that, whatsoever he [Raleigh] shall take upon him to say for me ... you will no more believe it. ... Would God I were as free from offence toward God in seeking, for private affection, to support a person whom most religious men do hold anathema.

Whether James had known of Raleigh before Howard's missive is likely, but not certain. That he knew of him now, and would regard him warily for as long as he reigned in England, was assured.

When James finally did replace his cousin, one of his first acts was to dismiss Raleigh as captain of the guards. Shortly afterward, he eliminated or sharply reduced monopoly income for all of his subjects. It is possible that neither act had Raleigh as its specific target. In the first case, the king might have been behaving like a modern politician who has won an office previously held by the opposition; he might, in other words, have been cleaning house, bringing in his own people. Even then it was standard procedure. In the second case, James seems to have made a decision according to his own vision of government, perhaps not even knowing who would be affected by it and to what degree.

Nonetheless, and to the surprise of no one who knew him, Raleigh took the two decisions personally. A third royal decision *was* personal: He was ordered to vacate Durham House, even though he had spent a great deal of money remodeling it and converting some of its rooms into a laboratory for experiments in medicine. The new king, Raleigh now felt certain, was out to get him, and he did not know why and could not get a satisfactory response from those he queried. What he *could* do was reveal his own displeasure, and that he did. Soon the mutual antipathy between subject and king would become a matter of public record, with James, in the *Counterblaste*, referring pointedly to Raleigh's efforts on behalf of tobacco. "It seems a miracle to me," he wrote, "how a custome springing from so vile a ground, and brought in by a father so generally hated, should be welcomed."

Raleigh's friends were alarmed. Some even began to fear for his safety if the king were not somehow appeased. A few urged Raleigh to speak to him, if not exactly to ingratiate himself, then at least to blunt the edge of hostility. Raleigh resisted, believing with some justice that he had done nothing wrong and it was James who should make a move. But that was impossible; rulers did not apologize to subjects and Raleigh knew it. So he finally relented, deciding to act, although in a manner so self-defeating that he might just as well have asked his highness how that tongue of his had gotten so big and unmanageable.

Raleigh bought the king a present. A book. It was called *A Discourse Touching a War with Spain and of the Protecting of the Netherlands.* He could not have chosen worse and had to have known it. Among other things, the volume advocated taking up arms against Spain, and James did not want to take up arms against Spain. It advocated a paternal relationship with the Netherlands, and James did not care about the Netherlands one way or another. He might not even have been able to find it on a map. The *Discourse* was not reading material; it was a slap in the face, a kick in the shin, an insult of the first degree. James took it as Raleigh's not-so-subtle way of letting him know that he, Raleigh, did not approve of the royal handling of affairs. Perhaps this is what Raleigh intended, to tweak the king rather than propitiate him. No other, more satisfactory explanation for the gift has ever been offered, although Raleigh's desire to provoke, if that is what it was, is not easily explained, either.

Sometime after that, with the air between the two men still highly charged, James's advisers got word of a plot against him. It was hatched by a certain

Lord Cobham and at least one agent of another country. The details need not concern us, except for the fact that Raleigh was believed to be one of the conspirators. Called before the Privy Council, the king's official and most highly placed advisers, he was asked about his involvement. He denied any and further said that he did not even know that a plot existed. If he had, Raleigh swore, he would have done his duty and reported it.

He was probably telling the truth, but James decided not to believe his adversary and, in fact, to use the incident as an excuse not merely to punish him but to send him to his death. He wanted no more of Raleigh's disloyalty, no more of his impertinence, no more of his existence.

Raleigh was startled, not to mention frightened; he had never expected things to come to such a pass. But he got his wits back about him and defended himself both eloquently and courageously.

James was unmoved. Others, however, looked at Raleigh as they had never done before. As Giles Milton puts it, "For years, Ralegh had been an object of hatred and ridicule among the poor and needy. Now, those very same people took pity on the underdog and made him the man of the hour." Said one of them: "Never was a man so hated and so popular in so short a time."

James was surprised by such a response, which was so great that his advisers told him he could not ignore it. He had been forced into leniency. Rather than beheading Raleigh, the king settled for returning him to the Tower of London. The first stay had lasted a month. This one would be twelve years.

It was not as bad as it sounds. The king seemed pleased enough simply to have his nemesis out of commission; he showed no desire to make him miserable in the process. The prisoner was able to take his family and servants with him to his place of incarceration, and Raleigh's son Carew was conceived while they so dwelled.

The lodgings were spacious. The prisoner and his brood occupied a large study and an equally large bedroom, and the ceilings had been raised prior to their arrival. They lived much as they would have lived had they still been free, perhaps even with less financial pressure. A few historians have gone so far as to describe Raleigh's years in the tower as comfortable. They were certainly productive. He read and drew maps, no doubt dreaming of routes he would take once free again. He continued his medical researches, at one point "macerating forty roots, seeds, and herbs in spirit, distilling the result, and then adding powdered bezoar stones (formed in the stomachs of animals), pearls, coral, deer's horn, amber, musk, antimony, and sugar." The result was a "great cordial," and Queen Anne of Denmark, sampling it some time later, claimed that it saved her life.

Raleigh also wrote his best-known book in the Tower. *The History of the World*, a volume no less ambitious than its title, "could be called the *Das Kapital* of the seventeenth century," says biographer Robert Lacey, "for just as Marx analyzed history in order to justify his thesis for proletarian revolution, so Walter Ralegh examined the past to illustrate the workings of God in political events—particularly in the punishment of unjust rulers."

James, apparently, did not see himself in this category.

IN 1616, SIR WALTER RALEIGH was released from the Tower of London to lead another search for New World riches. It was his idea. He had been lobbying to return for several years: dropping hints, telling friends who might pass the word to the king or those near him, petitioning the Privy Council, and writing letters, even a letter to the king's wife. But it was James's son, Prince Henry, who seems to have been responsible for the decision. The prince had read Raleigh's earlier book and agreed with him that in the large, rich, and beautiful empire of Guiana, and especially in the great and golden city of Manoa, there must be precious stones, sparkling gems, wealth untold—if only one knew where to look. Raleigh, with his natural ebullience fueled all the more by a zeal to escape captivity, said he did. He had found all the wrong places on his previous voyage; this time, he assured the prince, he would find the right ones, by process of elimination if nothing else.

He had better, young Henry told him. The journey was not a joyride, nor did anyone at the palace care about Raleigh's reputation. The exchequer needed money; that was what mattered, and only that. James had been overspending even by his own profligate standards, so much so that he was now willing to try anything—and *anyone*—to fill the royal coffers again. Raleigh was a last resort, the prince let it be known, not the recipient of a vote of confidence.

He was also a lost cause. Plagued by disease and foul weather, not to mention his own overselling of his prospects, Raleigh fared no better under James's banner than he had under Elizabeth's. Once again, he discovered no El Dorado, in Guiana or anywhere else. He unearthed no rare minerals, spied no collection of jewels or even baubles. He did manage to get his hands on a few gold ingots but only because he stole them from some Spaniards—and that was another problem; James had let Raleigh know prior to departure that under no circumstances was he to attack any Spanish settlements in the Americas.

In fact, Raleigh had disobeyed with a vengeance. He ransacked several villages and destroyed altogether the town of San Thome, where his older son Wat, the cherished child born out of wedlock, was killed. Raleigh's friend Lawrence Keymis, who had led the troops in battle and been more directly responsible for the violence than Raleigh himself, was so disturbed by the turn of events that he committed suicide, wedging a knife into his heart and dying almost instantly.

Raleigh was devastated. His boy was gone, his long-time comrade-in-arms was gone, his prospects for a future were gone. In retrospect, it almost seems as if Raleigh went to Guiana to act out a death wish of his own. Writing to his wife shortly afterward, he spoke not only of the tragedy that had just occurred but of the one that lurked. "I shall sorrow for us both," he said. "I shall sorrow the less, because I have not long to sorrow, because not long to live."

In October 1618, back in England, Sir Walter Raleigh went on trial for both his disobedience to the crown and his failure to enrich it materially. The official charge was treason. On the day of sentencing, he was escorted from his cell to stand before the judges who controlled his fate. "It was a humbling final journey," writes Lacey, "and Walter, still shaking from fevers he had contracted on his voyage to Guiana, looked a broken man." His hair was long and uncombed, his eyes cloudy, his voice hoarse.

His ears, however, were just fine. He heard the pronouncement of death, just as he expected. James ordered that it be carried out the next morning, Lord Mayor's Day, when there was a host of other activities in London. The king hoped that these would engage the citizenry, thereby ensuring a small turnout for Raleigh's demise. James wanted to end the life of his nemesis, not begin his martyrdom.

The rest of the day proceeded slowly for the condemned man. We do not know what he did, how well he slept, or whether he slept at all. He handed down none of his thoughts to posterity. In the morning, a few minutes before being led to the scaffold, he asked for his pipe, slowly packing it with tobacco and taking the time to savor it.

This last pipe of Walter Raleigh's now resides in London's Wallace Collection and bears the inscription, "It was my companion in that most wretched time." One wonders whether there was a melancholy smile on his lips as he drew on it, to what extent his senses were attuned to a taste he would never know again, to feelings he would feel for the final time. One wonders how much comfort he felt as the smoke encircled him with its familiar, wavering presence. And one wonders, when Raleigh took his last puff of his last pipe, whether he mused on the possibility that the embers in the bowl would outlast the man who had set them afire.

John Aubrey, who was apparently there during the final hours, said that "some female persons were scandalised at" Raleigh's lighting up so soon before meeting his maker. Aubrey, however, understood. He thought " 'twas well and properly donne to settle his spirits."

But the scaffold, that great *un*settler of spirits, awaited. Raleigh was taken by guardsmen to the steps and ordered to begin climbing. As he did, his boots thudding on the wooden planks, he lifted his head, looking around to see a huge crowd before him and behind him and off to both sides, people from all stations of English society, standing shoulder to shoulder in eerie and expectant silence. Some seemed to nod in his direction. Others averted their eyes. Still others bowed their heads as if in prayer. In overwhelming numbers, London had come to pay its respects to Sir Walter; there would be another Lord Mayor's Day next year.

Raleigh took off his gown and doublet and asked the executioner to show him his ax. He touched the blade, tapping it a time or two with his finger. "This is a sharp medicine," he said with a wan smile, "but it is a sure cure for all diseases."

He removed his hand from the weapon, and the executioner offered him a blindfold. Raleigh turned it down. Then the executioner spread his gown on the scaffold, inviting Raleigh to kneel on it. He did, said thank you in a soft voice, and closed his eyes and waited as calmly as he could for a violent death.

Several moments passed. If possible, the throng around him grew even more silent, the air more thick with dread. Raleigh felt nothing more than a few unthreatening breezes on the back of his neck. He opened his eyes. The executioner was looking back at him, not yet having moved his ax into the ready position. Raleigh lost patience. "What dost thou fear?" he said, voice booming now. "Strike, man, strike!" And he turned away.

The executioner struck. Raleigh's head was severed from his body by the second blow. The crowd gasped; people were horrified even though they all knew why they were there and many had witnessed executions in the past. But this was Sir Walter Raleigh's execution, Sir Walter Raleigh's blood on the scaffold, Sir Walter Raleigh's head in the basket—and who could truly believe it until it had happened?

The gasps rode out on the breezes, and they were the only sounds to be heard. One man broke the spell after a minute or so, calling out, "We have not another such head to be cut off." Several people murmured in agreement, then all began, slowly and disconsolately, to depart.

As for the king who had ordered the decapitation, he had seven years of life left to him. He would be tormented by tobacco through all of them, and in ways he could never have foreseen.

THREE

The Politics of Tobacco

THE FIRST PERMANENT colony of English citizens in the New World was named for him, but in the beginning this seemed more of an indignity than an honor. James seldom paid attention to Jamestown, and when he did it was usually to complain. He had enough difficulties at home. Why had he sanctioned a whole new set of them so far beyond his reach? Why were those difficulties so resistant to change? Was there anything he could do, other than abandon the whole enterprise, to calm his nerves and restore his pride?

He had told the people who founded the settlement "to preach and baptize into the *Christian Religion* and by propagation of the Gospell, to recover out of the arms of the Divell, a number of poore and miserable soules, wrapt up into death, in almost invincible ignorance." But the settlers, an unimpressive lot on their best days, could not meet such a challenge. They were barely able to get across the ocean, and once they did they seemed fresh out of initiative.

Some of them were gentlemen. The title was supposed to reflect a certain refinement of birth and style, perhaps even a degree of character. "But by Elizabeth I's time," writes J. C. Furnas, it "was accorded pretty much anybody with a clean shirt and money in his purse that he had not earned with his own hands."

Others relocating to Jamestown were even less distinguished, a collection of "oddsticks who had not been very successfully at anything in England." As a publication of the time called *The New Britannia* put it:

> Our land, abounding with swarms of idle persons, which having no means of labor to relieve their misery, do likewise swarm in lewd and naughtie practices, so that if we seek not some ways for their foreign employment,

> we must supply shortly more prisons and corrections for their bad connections.

The rest of the passengers on the three ships that first landed in Jamestown were "Tradesmen, Serving-men, libertines, and such like, ten times more fit to spoyle a Commonwealth than either begin one or but helpe to maintain one." There were two women on board, one of the "gentle" variety, the other her servant. All in all, it was the perfect cast of characters for an inauspicious start, and that is exactly what they produced.

But the inappropriateness of the location caused as many problems as the incompetence of the colonists. Jamestown was founded on a small peninsula in Virginia, about sixty miles from Cape Henry. It was close to the ocean, providing ease of both trade and transportation, but bereft of virtually all other advantages. The land had for years been home to "miasmic vapours"; it was bug-infested, swampy, and many miles from a source of fresh water. As a result, scurvy, malaria, and dysentery tormented the new arrivals even more than the native tribes, who, after an initial period of relatively peaceful coexistence, decided that they wanted the land to themselves again and sometimes expressed their feelings harshly. Of the 104 people who landed in Jamestown in the spring of 1607, forty-six did not survive the summer.

The winter was even more of a trial. Since there were few if any farmers among the original settlers, it had not occurred to more than a handful of them to plant crops when they arrived. Those that *were* planted were quickly harvested and eaten, and almost all of the victuals that had been brought on the ships were consumed with equal haste. In desperation, the Jamestowners began eating their hens and sheep and horses, satisfying their appetites for the moment but reducing their chances for successful farming, and for a successful colony, even further. When their animals were gone, they turned in greater desperation to the stray creatures of the wild: "doggs Catts Ratts and myce."

And so it was that even Jamestown's first governor, Thomas Dale, a man with a powerful allegiance to projects of the crown, could not bring himself to speak optimistically about the settlement's chances. "Every man almost laments himself of being here," Dale admitted, and when the winter of 1609–10 came along in all its fury, he could easily have dropped the "almost."

It was called "the starving time," and it is impossible to imagine a group of human beings suffering more at the hands of the providence in which they so fervently believed. According to some estimates, the population of Jamestown stood at five hundred in the autumn of 1609. By winter's end, there were a mere sixty men and women in the colony, the result of "disease, sickness, Indians' arrows, and malnutrition." And, it should be added, desertion, as some of those who did not succumb to the preceding misfortunes fled from Jamestown as fast as they could, disappearing into the forest and never again showing themselves to their countrymen.

But it was not just with arrows that the Indians showed their disapproval of the newcomers. They also withheld corn and other foodstuffs, in addition to seedlings, upon all of which the settlers had come to depend. They were no longer willing to barter, no matter what the white man offered in return, and on occasion they were not even willing to communicate. Among other things, this caused the Jamestowners to fight among themselves, apportioning blame for the hunger, dissension now as much a threat to their survival as adversity.

Captain John Smith, one of the colony's few able residents, told a story that is probably true despite the lengths to which it stretches credulity. It seems that one of his fellow settlers was so famished he killed his wife, "then ripped the childe outt of her woambe and threw itt into the river and after chopped the mother in pieces." Then he cooked her. Or perhaps he had already cooked her or did not cook her at all—that part is not clear. But at some point after he was done hacking her to bits, he sprinkled her remains with salt, which was also known as the time as powder. He grabbed a knife and fork, dug in. Once his cherished spouse, the poor woman had now become his dinner.

The man was caught after only a few bites, however, and summarily put to death. It was a ghastly incident, perhaps the most horrible experience in the lives of those who knew either perpetrator or victim. Yet musing on it a few months later, Smith could not avoid a certain wryness. "Now whether she was better roasted, boyled, or carbonado'd, I know not," he observed, "but of such a dish as powdered wife I never heard of."

Others dined that winter with equal disregard for niceties, although not with equal disdain for marriage vows. They ate "boots, shoes, or any other leather, and were 'glad to make shift with vermin.'"

The weather only added to the torment. There were blinding snowfalls, huge drifts, and winds that cut into the skin like the tips of the natives' arrows. The colonists had never seen anything like it in England. Nor had they ever felt temperatures this low. A few people chopped down their cabins, burned the wood for heat, and then moved in with friends. Those who survived "the starving time" would remember it as the most harrowing few months of their lives, a winter when

> despair moaned on the pestilent winds that swept across the surrounding swamps; fear grew into an oozing thing, like the brackish water from the river seeping into the well; even the imagination of a Milton or a

Dante could not conceive of a hell worse than these dispirited wretched people faced.

IN THE LONG RUN, though, there was a more serious concern for the colony than any it faced that winter. Put simply, Jamestowners had no way to support themselves. That they did not grow enough of their own fruits and vegetables and grains was only part of it. More to the point, they could not find anything, edible or otherwise, that was suitable for export. They tried olives, pitch, timber, tar, sassafras, soap ashes, and cedar, packing them up and shipping them off to the Motherland in hopes that they would find a market.

But England was not interested. Either it did not care for the colony's products or it could get a higher grade elsewhere at a competitive price. There was no sense of duty here, no show of kinship between mainland and outpost. The English would not support their New World brethren unless their goods could succeed on their own, and that was something, in the early years of the colony, they just could not do.

The colonists kept plugging away. They experimented with glassblowing, at which they failed because of faulty equipment, and then silk making, at which they were encouraged by that noted "silkworm buff," James I. He thought this a noble pursuit for his countrymen abroad, all the more so because he dabbled in it himself. In what must have struck him as a gesture of great magnanimity, he even sent some of his own, personally bred worms to Jamestown, hoping they would thrive on the mulberry trees that grew plentifully around the settlement.

They did not. They might have been damaged in transit; they might have been repelled by the climate; they might simply have been a batch of wiggly little invertebrates whose best days were behind them. Regardless, they no sooner arrived in North America than they dropped dead, not having added so much as a single square inch to the world's stores of elegant fabric. Actually, a few of them did not die immediately; their fate was to be eaten by rats. North America did not treat the insects from abroad any more kindly than it had the humans.

James next urged the colonists to raise grain. At this, he believed, even the societal rejects who inhabited Jamestown could not fail. But Captain Smith was opposed. He admitted that the Virginia soil was conducive to corn; the natives had been growing it for centuries, and the colonists had begun to plant a few crops of their own and were improving with each harvest. But corn would not make a good export, Smith insisted. It would not sufficiently reward the amount of labor required, especially given the complicated set of price controls that the crown had imposed and would not consider revoking. So it was that, with no little trepidation, Smith tried to persuade the author of the *Counterblaste to Tobacco* to consider that which he had counterblasted.

The settlers had been growing an indigenous strain of leaf called *Nicotiana rustica* for a year or two now. *Rustica*, however, was "poor and weake and of a biting taste." The natives liked it well enough, but the more refined English palate, although it would settle for *rustica* if it had to, preferred a milder type of weed known as *Nicotiana tabacum*. But *tabacum* did not grow in America. Nor did it grow in the England. Rather, it sprouted in greatest profusion from the soil of the West Indies, and therein lay a dilemma not just for the colony but for the crown, as well.

The West Indies were controlled by Spain, as they had been since Columbus's time. The Spaniards sold their West Indian *tabacum* to England and were eager to sell more, as much as they could, but only at exorbitant prices. They charged as much as six times what the colonists asked for their own, inferior brand of leaf. James raged at the expense, which only made tobacco more of a vexation to him than ever and perhaps led him to think that war with the despised Spanish might not be such a bad idea after all.

Short of formal hostilities, though, the king had no choice but to pay. His subjects, quite simply, would have rebelled had he followed his heart and either ended the tobacco trade or turned to North American *rustica*. Damning them for their addiction as much as he did the Spanish for their greed, he kept the tobacco coming and roiled at the irony of his plight: Jamestown struggling for its very life while the sorely depleted exchequer coughed up as much as 200,000 pounds sterling a year to the long-time bête noire of the British for a product that was reviled down to the marrow of his bones by the very monarch who authorized the payments. It was an intolerable situation for James, and not until a young man named John Rolfe arrived in the New World at the end of "the starving time" did a way out begin to appear.

Rolfe is best known to history for his marriage to the native American princess Pocahontas, which brought occasional periods of truce to the otherwise stormy relations between the various tribes of Virginia and the English colonists. Prior to meeting Pocahontas, Rolfe had spent a year shipwrecked in Bermuda, hungry and diseased and not knowing whether he would survive. Perhaps traumatized by the experience, or perhaps simply needing some time to recover, he struck those who first met him in Jamestown as a shiftless sort, largely absent of purpose. Unable to figure out what else to do with himself, he began to experiment with tobacco.

He did so for several years, and things did not start out promisingly. Rolfe could do nothing to domesticate the unruly *rustica.* It continued to taste bitter, draw unevenly, and resist all efforts at improvement. To some settlers, the leaf even had a gritty touch. It was not something they wanted to hold in their hands, much less stuff into a pipe or form into a cigar and inhale.

Rolfe was just about to give up, to find something else to keep him occupied, when he got an idea. He persuaded a friend who was a shipmaster bound for Trinidad to get hold of some *tabacum* seeds and bring them back with him to Jamestown. Rolfe would conduct a final round of trials on this type of leaf, not knowing whether it would adapt to the local climate and soil, but also not knowing what else to try. If he failed with the Trinidadian product, he would forget about the weed once and for all.

The seeds arrived late in 1611. The colonists planted them as soon as they could but were disappointed with the results. The *tabacum* made for better smoking than the *rustica,* but it was not as flavorful as the *tabacum* that grew in the Indies and was so venerated in England. *Was* it the climate? The soil? Maybe there was a problem with the way had Rolfe cured the leaves, which is to say, the way he dried the sap from them. Or maybe it was the amount of time he had allowed for curing, or the process of shredding the tobacco to prepare it for smoking. There were so many opportunities for error; so few chances to discover it.

Back Rolfe went to the drawing board. Or the curing house. As Carl Ehwa Jr. has pointed out, tobacco "becomes a source of pleasure only after careful implementation of relatively complex growing and procedures." It is "dependent upon the nurturing of variable properties of structure, color, and chemical composition for its final commercial appeal." It was with the nurturing of these variables that Rolfe struggled.

But he was determined, unflaggingly so. He kept fiddling, modifying, adjusting. Most of his fellow colonists were discouraged, but he would not join them. He rallied his mates, cheering them on and bucking them up and at one point writing, "Tobacco [is] verie commodious, which thriveth so well that (no doubt) after a little more triall and experience in the curing thereof it will compare with the best in the West Indies."

No one believed him. There were probably times when he did not even believe himself. That he turned out to be right, says the perceptive observer P. A. Bruce, "was by far the most momentous fact in the history of Virginia in the seventeenth century."

In 1615 and 1616, the Americans shipped a total of 2,300 pounds of tobacco to England. The next year, after Rolfe had solved most of his problems, the colonists sent 20,000 pounds abroad. Preferring the Virginia crop by a vast margin to their own, and even rating it the equal in many ways of the West Indian strain, the English promptly consumed every leaf that the colony provided and demanded more, as much as it could grow.

Prices for the American leaf shot up. The importance of Spain to British smokers shot down. In 1619, Jamestown exported twice as much tobacco to the Motherland as it had just two years earlier, and after another couple of decades its annual exports would exceed a million and a half pounds. The once benighted settlement in the New World swamp had not merely discovered how to survive; it was prospering, splitting at the seams now with improbable productivity and pride, turning into the first boom town in a country that would eventually give birth to dozens of them, hundreds, centers of frenetic action, giddy optimism, and sudden and prodigious fortunes. The magic that John Rolfe had so diligently worked with the tobacco seeds from Trinidad was responsible for "this nation's first business," and capitalism was off to a rousing start.

But the leaf's contributions to Jamestown were exceeded by the even greater contributions it made to the entire land that would one day be the United States. For James had lost patience with his colony. The great silkworm debacle seems to have been the last straw for him. He wanted no more aggravation, no more drain on his finances. Had Rolfe not provided his fellow settlers with an economic base precisely when he did, especially one as lucrative as tobacco turned out to be, the king might have written off his eponymous village, allowing it to die and the colonists to scatter themselves through the forests, eking out an existence like animals or tribesmen, or else to return, abashed, to England to start anew from scratch. If this had happened, James more than likely would not have tried to colonize this benighted part of the world again, and it might have been a while before any of his successors made an attempt of their own.

It is possible, then, that another nation would have filled the void, that Virginia might have had French roots or Spanish roots or the roots of some other European empire. If so, it would have been a different part of America from what it eventually became, and the country itself, as a result, would have been a different place, although in what ways it is impossible to guess. On more than one occasion in the seventeenth and eighteenth centuries, tobacco played a major role in defining the United States that we know today. John Rolfe's transformation of tobacco, from bitter to mild, from burden to boon, was the first of them.

JAMESTOWN DID NOT handle its success well. The colonists were more than merely thankful for tobacco; they were bewitched by the affluence it brought, so much so that they soon allowed themselves to become enslaved by it. Perhaps they had no choice. The bad times were but a few years past, still such vivid and painful memories; Americans, in the jubilation of their relief, could not refrain from making tobacco the sum total of their society, the preoccupation of all their waking hours. It followed that "virtually everything else was neglected," writes Ivor Noel Hume, "including their spiritual life."

Their agricultural life was the more pressing difficulty. Governor Dale, who only a handful of years before had admitted that everyone in Jamestown wanted to be somewhere else, now lamented the fact that those who had stayed could think of nothing but the rich, brown leaves of the tobacco plant. He told them to grow other crops in addition, to *care* about other crops, "since they could not eat money." When they balked, he tried to legislate good sense, a proposition with a long history of failure. On one occasion he ordered that only a single acre of Jamestown be devoted to the weed for every two of corn. A later edict limited settlers to 1,000 tobacco plants each; in time, the number was raised to 3,000.

Nobody cared. Nobody listened. Nobody even acknowledged that laws were being passed and that at least a few people expected them to be obeyed. Tobacco had become the mistress of men's hearts, and passion cannot be regulated by statute. When Jamestowners were not selling the leaf, they were planting or harvesting, watering or pruning, curing or shredding—and they were lighting up and sucking in and blowing out through every single day of it.

In 1617, Captain Samuel Argall replaced Dale as governor of Virginia. He was startled by the toll the weed had taken on Jamestown. The economy was still thriving; the colonists were exuberant, and the pace was as hectic as it had been for the past decade, ever since tobacco had become the mother lode. But that was what disturbed Argall so much. The pace was *too* hectic and, it seemed to him, driven not by fulfillment or even duty but by despair, the fear that if anyone slowed down so much as a step, planted one less seed, or took one extra day in the curing, there would be a new starving time. It is those who have suffered the most who most dread suffering and who go to the greatest lengths to avoid repetition, regardless of whether their actions are rational.

What surprised Argall more than anything else was the fact that, in some ways, the boom town had taken on the appearance of its opposite number, a ghost town. The United States would spawn a number of these—but Jamestown? Somehow, triumph seemed to have wreaked the same havoc as tribulation; joy had brought about the same neglect, the same lack of pride in surroundings, as misery. On first arriving in the tobacco capital, Argall reportedly found

> but five or six houses, the church downe, the palizado's broken, the bride in pieces, the well of fresh water spoiled; the store-house … used for the church; the marketplace and streets, and all other spare places planted with tobacco … the Colonie dispersed all about planting tobacco.

Two years later, Argall was himself replaced as governor of Virginia. Officials in London encouraged the new man, Sir George Yeardley, to do something about Jamestown's fixation with the weed:

> We have with great joy understood of your arrival in Virginia, and of your firm resolution to reforme those errors which have formerly been committed. One chiefe whereof hath been the excessive applying of tobacco, and the neglect to plant corne which of all other things is most necessaries for the increase of that plantation. Wee therefore … earnestly pray you that nothing whatsoever may divert you from that worthy course.

Yeardley was not diverted. He tried his best to reform the errors. But the citizens of Jamestown would not be diverted, either, and it was their priorities that prevailed.

As a result, some people thought the colony had sold its soul to the devil and that the horned one had made a remarkably canny transaction. After all, he had given the colonists everything they wanted except the ability to use it wisely, and what could be more damning than that? It was just the kind of game the devil liked to play, this demonstration that wealth could be a form of bondage no less than poverty, just the kind of lesson he liked to teach, especially if he could put it across the hard way. He must have been smiling down there where he lived, accepting the cheers of his minions in that place where smoke was even more plentiful than in the colony of Virginia.

A few later observers saw evidence of the devil, or something like it, in the way Americans behaved toward one another, claiming that tobacco had been the death of brotherhood. "The society that was designed to be a productive and diversified settlement in the wilderness," write James West Davidson and Mark Hamilton Lytle, "soon developed in a world where the single-minded pursuit of one crop, tobacco, made life as nasty, brutish and short as anywhere in the hemisphere."

This, however, is nonsense—history with an agenda, the kernel of truth at its core almost unrecognizable in the husk of exaggeration. Most of the hemisphere was less civilized than Jamestown, far less civilized. And most of the less civilized places were nastier and more brutish. But even John Rolfe had to confess that his fellow citizens were not adjusting smoothly to their new lives, that, at the very least, they had lost their sense of proportion. "Wee found the Colony (God be thanked) in good estate," Rolfe wrote to England during Samuel Argall's governorship, "however in buildings, ruined fortyfications, and of boats, much ruined and greate wante."

But the devil behind tobacco was not a simple sort. What he had in mind was something far more insidious than just ruined boats, broken bridges, and other signs of physical disrepair. Tobacco is a crop with a voracious appetite for the nutrients in soil, particularly nitrogen and calcium and potash, and it was all the more voracious in Virginia because the topsoil was so thin, not having been enriched by glacial mineral deposits. "It was," thus, as Daniel Boorstin explains, "only on virgin land that tobacco could flourish; the second crop was usually the best."

By the time the third crop came along and was ready for harvesting, the land was beginning to wear out. Every two or three years, the colonists had to give their existing acreage some time off, either let it lie fallow or plant it with corn or wheat, and this meant finding new acreage to replace it. It meant hacking new fields out of the wilderness for the fourth crop. It meant chopping down trees and clearing vegetation and removing animals. It has been estimated that, largely because of tobacco, half a million acres of Virginia were deforested by the end of the seventeenth century. Kirkpatrick Sale, no friend of the white man in any of his colonial endeavors, finds the consequences of such farming varied and destructive:

> Extermination of the beaver, for example, meant the deterioration of beaver dams and ponds and the unchecked flow of stream waters, which destroyed riverine eco-systems, increased floods and topsoil runoff, promoted bank erosion and siltation, and reduced water tables.

Another harmful effect of the leaf was what it did to relations between Englishman and native. Never good to begin with, they worsened dramatically with the white man's cultivation of tobacco. For it was not just plant and animal life that had to be removed from the land that the colonists so desperately needed; it was the humans who had been living there for centuries. As the white man's share of the New World expanded for his crops, the natives' share shrank the same amount. The natives were constantly being pushed back, farther and farther inland, farther and farther from their homes and familiar settings, to which they were attached for spiritual reasons as well as for convenience and familiarity. Where once the native tribesman had made a life, there now grew the aromatic leaf that seemed even more important to the interlopers than it had been to those who knew it first.

Finally in 1622, with Pocahontas five years in her grave and the tenuous peace she had helped to forge having departed with her, the tribe of which she was born, the Powhatans, took revenge, killing 350 Americans in a village near Jamestown. One of the victims might even have been John Rolfe, who died at the approximate time of the attack, although whether of illness or "a tomahawk or feathered arrow" is not certain. He was thirty-seven. Surviving him were his second wife and two children, one by Pocahontas.

The colonists struck back quickly, themselves murdering and burning and plundering. The violence escalated, and then the escalation escalated. Both sides behaved with vicious indecency, making victims of male and female, young and old, warrior and innocent alike. But the natives were no match for the Americans in terms of weaponry or military maneuvering. They fought with enthusiasm more than skill, with recklessness instead of calculation. They did not have a chance. By the 1640s, many of the Virginia natives had been wiped out altogether, a black mark on the white man's soul. Those who survived did so by fleeing or, if they remained, by offering little in the way of resistance. Most had lost their leaders, their nobler and more astute members, and were made up now to an unfortunate degree of beggars, dependent on the colonists' charity, of which there was little, and his tobacco, of which supplies were becoming more and more plentiful all the time.

JAMESTOWN AND THE CROWN were not getting along much better than Jamestown and the American tribes. Finally the spot on the map that bore the king's name had found a way to earn its keep, but to what did it owe the turnabout? Tobacco. Finally the English had discovered gold in the New World, but what color was it? Brown. Finally the English were able to reap some benefits from a colonial undertaking, but who was responsible? Sir Walter Raleigh—or so it might have seemed to James in his more tortured moments, of which, as the years went by, there were many. Ambivalence burned like a fever in him, and his machinations at the time reveal the agitated state of his emotions.

First he decided that growing tobacco in England, always a dubious venture and never anything more than a sidelight to the economy, was "to misuse and misemploy the soill of the kingdom"; there would, henceforth, be no more of it. Then he ruled that purchasing the weed from Spain was no longer necessary; this would cease, as well. From now on, all tobacco smoked by Englishmen would be provided by Jamestown, and all Jamestown tobacco—at least, all that was not consumed by the colonists themselves—would be shipped to England and to no other markets:

> Whereas it is agreed on all sides that the Tobacco of these plantations of Virginia ... which is the only present means of their subsisting, cannot be managed for the good of the Plantations unless it be brought into one hand, whereby the foreign Tobacco may be carefully kept out, and the Tobacco of those plantations may yield a certain and ready price to the owners thereof, That to avoid all differences and contrarieties of opinions ... we are resolved to take the same into Our Own Hands.

Furthermore, the crown would set the prices, and they would be far more favorable to smoker than to grower. James also insisted that all costs related to shipping—freight, insurance, handling, inspection, storage—be paid by the Americans. This, as more than one historian has noted, amounted to the New World's first tax. Raleigh might not have discovered riches on his trips to Guiana, but the king now seemed to have found a way to replenish the exchequer without them.

James probably discussed these measures with his advisers. There were at least a few prudent men among them, willing to speak their minds in most circumstances—but did they in these? Did they tell James how the colonists were certain to react? Did they assume that the king knew the answer but did not care, that he had no qualms about riling them? Or did they hold their tongues in this case, deciding that James needed to let off steam more than he needed to govern wisely? There are no answers, only the speculation.

What *is* known is that the citizens of Jamestown were stunned. At first they wondered whether there had been a mistake, a miscommunication of some sort. James had granted a charter to English citizens of the New World, assuring them of "all Liberties, Franchises and Immunities ... as if they had been abiding and born within this our realm of England." Was this the way the king interpreted the charter?

Through the crown's agents in Virginia the settlers replied. They did not believe they should be required to send all their exports to England; they wanted to reserve a portion for other countries, which would help to advertise the quality of their leaf, open even more markets for them, and stimulate higher prices. Those that James had decreed were so low as to be ludicrous; worse than ludicrous, actually—unprofitable. And his ruling that Americans must assume all shipping costs was so unfair as to be unconscionable. The English merchants who received the tobacco should split the costs evenly with them, some colonists thought; others said the merchants should pay more than half, as they stood to make more than half the profits.

James would have none of it. His position was that, in providing vessels for the tobacco trade and the capital to get it started, as well as the facilities for storage and sorting in England, the merchants were taking more of a risk than the Americans, who were only chipping in some leaves. If a ship were wrecked at sea or a warehouse looted or burned, the merchants could not replace their investment or could do so only at great expense. But if the settlers lost their tobacco, they could simply scatter some more seeds on the ground and grow a new batch. Nothing to it—plant and water.

Perhaps more important, though, James believed that, the charter notwithstanding, the New World existed primarily to service the Old—that it was, in effect, a wholly owned subsidiary. It was not to make its own decisions any more than Norwich was, or Manchester or Leeds or Brighton. For the Americans to think otherwise was for them to reveal something of which James had been wary ever since the first English colonists had landed in Virginia: that they would become spoiled by freedom and unmindful of where their responsibilities truly lay. "As it is atheism and blasphemy to dispute what God can do," the king said on one occasion, getting his tongue out of the way of his arrogance, "so it is presumption and high contempt in a subject to dispute what a king can do, or say that a king cannot do this or that."

What happened next only confirmed James's fears. America began to rumble. Colonists met in small groups at private hours and talked openly of their discontent with the monarch and his court, something they had never done before. Then they got practical, making plans, with different people willing to go to different lengths to vent their frustration. Some would bribe customs officials to allow them to sell their leaf in places other than England. Some would sail their ships the long way around officials who would not take the bribes. Some would engage in "socking," a process whereby men who worked on the docks would "plunge their hands into the bales of tobacco, sock away as much as they can hold and then resell on their own." Some would hide portions of their crops from royal inspectors and offer them instead to a "Smuggler's Fleet" of higher bidders, a fleet that would continue to do business with American growers for the rest of the century. As late as 1692, the collector of customs for the Chesapeake Bay would tell his superiors in London that "in these three years last past there has not been above five ships trading legally in all those rivers and nigh thirty Sayle of Scotch, Irish and New Englandman."

Yet other Americans would continue to forward all of their yield to the Motherland, but, in the manner of Prohibition-era bootleggers who either watered or poisoned their hooch, they would increase the amount of the shipments by adulterating them, mixing their tobacco with salt, straw, dirt, ground glass, paper, garbage and sometimes even the fecal matter of animals. No one seemed to notice. As Shah Abbas could have told James, a man who craves his leaf is not a man to make fine distinctions.

Even so, England profited mightily from American tobacco. "In the 1660s," writes Jordan Goodman, "for example, tobacco duties from the Chesapeake colonies accounted for roughly one-quarter of total English customs revenues, and as much as 5 per cent of total government income."

But soon there were uprisings in colonies other than Virginia, which were now dependent on tobacco themselves. In Maryland, the leaf became "practically the do-all and end-all of early agriculture." It gave South Carolina "soe good a reputation ... that it will prove the most flourishing plantation that was ever settled in America." Pennsylvania "became renowned for its tobacco," or at least Peter Stuyvesant thought so in 1656.

Three and a half decades earlier, when William Bradford landed in Plymouth, Massachusetts, he found tracts of land that were "for the most part a brackish and deep mould much like that where groweth the best tobacco in Virginia." The Puritans started growing their own. So did their later neighbors to the south, where the Connecticut and Farmington river valleys soon became overrun with the weed, and where a law was passed in 1641 declaring that people who lived in the valleys could smoke only the tobacco that was grown in the valleys.

Tobacco farmers in Rhode Island, seeming as infatuated as the early Jamestowners, had to be reminded not to neglect other crops. When Sir Edwin Sandys, a member of Parliament, urged them to pay "greater attention to hemp, silk and the grapevine," they told him it was out of the question; tobacco demanded too much of their time and energy, was too important to their lives.

And so it was inevitable that Marylanders would join Virginians in bribing the customs brigades; that growers in Connecticut and Massachusetts would sail their ships on circuitous paths away from inspectors; that South Carolinians would contact smugglers; and that Rhode Islanders and Pennsylvanians would adulterate their crops. Ill will became as much a part of the American culture as tobacco smoke.

These acts of rebellion naturally incensed the rulers of England, from James I to George III. They were the first disagreements of note between parent and child. In time, of course, there would be others. The colonists would object to the levies that the crown wanted them to pay for defense against Indian attacks. They would object to export duties, on tobacco and other crops, of some 300,000 pounds a year. They would object, and would dramatize their objections, to the crown's attempts to impose a monopoly on tea. They would object to the crown's increased duty on molasses, which raised the price of the colonies' indispensable beverage, rum. They would object to the crown's demand that they provide living quarters, candles, beverages, and transportation for British troops stationed in America.

Perhaps if the colonists and their rulers had been able to work out a procedure to settle their differences about tobacco, they could have used the same procedure to settle—or, at least, soften—the later disputes. But they did not. Prices for the leaf continued to be a point of contention between New World and Old until they finally went to war. At that time, a tobacco farmer named Thomas Jefferson was almost 10,000 pounds sterling in debt to importers in London and Glasgow and complained that "these debts had become hereditary, from father to son, for many generations, so that the planters were a species of property annexed to certain Mercantile houses." The tobacco farmer Patrick Henry also found himself behind in payments and raged about the injustice of it. The tobacco farmer George Washington estimated that "in four years out of five his produce, shipped on consignment, brought lower prices than were quoted in the home market." Going back three decades before the war, says Forest McDonald, colonial planters did not even receive "a reasonable fraction of the wholesale price that their tobacco fetched in foreign markets."

And so the bitterness that grew out of Jamestown's salvation was allowed to become a foundation upon which future differences would seem all the more perplexing and, ultimately, unresolvable by peaceful means.

FOUR

The Rise of Tobacco

AS THE YEARS went by and Virginia turned into a more stable and civilized place, and as new colonies were founded and they, too, smoothed away their rougher, less hospitable edges, tobacco transcended its genre. No longer just a product for which people paid money, it became a form of money itself. The native tribes had long used a variety of items for this purpose: animal skins, shells, fishhooks, pots, salt, tools, and pieces of jewelry among them. To this list the European settlers of North America now added what some were calling the "Golden Token."

Because the price of tobacco, like that of all crops, fluctuated from colony to colony and season to season, it was not the most reliable of currencies. But given the unreliability of other currencies of the time, and the extent to which the various colonial economies depended on barter, which was often imprecise and more often inconvenient, tobacco soon became accepted almost everywhere. It was plentiful, enjoyed by virtually one and all, and credited with the very fact of Jamestown's continued existence. Why not spend the stuff as well as smoke it?

In 1619, according to John Rolfe, "a Dutch man of warre" arrived in Jamestown and "sold us twenty Negars." The medium of exchange was tobacco. Two years later, Virginians handed over 120 pounds of leaf for eleven maidens and a widow who had been conveyed to the colony for marriage to a dozen upstanding gentlemen. It was regarded as a fair transaction by all, with none of the bridegrooms finding their mates unsatisfactory, and several stating that "there hath been especial care in the choice of them … not any one of them hath been received but upon good commendations."

As for the wives, they seemed content to have elicited ten pounds of tobacco apiece, even though a hearty meal could be had in a tavern for precisely the same fee. A gallon of beer went for eight pounds. Other foods and beverages were also available for the weed, as were fabrics for clothing, implements for farming, and the occasional volume of verse or religious material. In fact, virtually anything that could be purchased for coin of the realm could also, at one time or another, be obtained for the right amount of sweet-tasting, easy-drawing, all beneficent American leaf.

It could also be used by an individual to pay his taxes. The Virginia Assembly accepted tobacco without cavil until 1633, when, for reasons not entirely clear, it changed its mind. No more of the tasty brown crop; from now on, cash and cash only would be accepted. Initially, people were upset with ruling and tried to think of ways around it. Then they decided that anger was a waste of energy. Better, they thought, simply to ignore the new law—and so they continued to pay their taxes with the yields of their acreage, even as the Assembly, despite having officially proclaimed that such compensation was no longer acceptable, continued to accept it.

Nine years later, the law prohibiting tax payment in leaf was repealed. It probably remained on the books as long as it did because no one could be bothered to remove it.

In 1673, a tailor named James Bullock, who lived in York County, Virginia, made a bet with his neighbor, one Matthew Slader. My horse can run faster than your horse, Bullock said. Slader disagreed and took the wager. But when Bullock's horse won an easy victory, Slader refused to pay. Bullock tried everything he could think of to get his money: pleading with Slader, bullying, threatening, even trying to humiliate him by telling Slader's neighbors that the man was a welcher. Finally, Bullock took him to court, certain that the scales of justice would tip his way.

They did not. The judge, sounding every bit as magisterial as a member of the House of Lords, ruled that it was "contrary to Law for a Laborer to make a race, being a sport only for Gentlemen." As a result, James Bullock, winner of the bet, was fined one hundred pounds of tobacco "for his insolence."

Larger amounts were assessed for more serious offenses, such as the transporting of Quakers from England, where few people liked them, to America, where few people wanted them. "Every master of a ship or vessel, that shall bring in any quakers to reside here after the 1st of July next," said a Virginia "Blue Law" of 1663, "shall be fined 5,000 pounds of tobacco." The same sum was charged to men and women of the colony who welcomed Quakers into their homes, especially if the Quakers intended "to preach or teach."

In Virginia, in the years leading up to the Revolutionary War, members of the militia were paid in tobacco, with leaves being carefully counted out and apportioned according to rank and experience. Most of the soldiers, being farmers themselves when not bearing arms, examined their compensation closely, wanting only plants of the highest quality for their labors. They had few quarrels with the amount of their pay, but it was not uncommon for them to complain about the color or scent or brittleness. Sometimes, depending on the status of their own particular crops, they might negotiate for seed instead of leaf.

Later, after the war began, the colonists purchased arms from Europe with tobacco, as well as ammunition and various supplies. Still later, they shipped the commodity overseas to pay some of the debts that they had incurred in the fighting.

Tobacco was also accepted as collateral for loans and as wages for midwives and mechanics and a number of other workers whom we would today identify as blue collar. A fellow could retire a debt with the weed; he could bribe a government official or anyone else in an official capacity who was willing to look in one direction instead of another and enjoy a pipe full of fine leaf as a reward. Early in the nineteenth century, Lewis and Clark made it "their principal token of exchange with the Indians" of the Louisiana Territory.

But perhaps most commonly, and without question most controversially, tobacco was the means by which churchgoers paid their preachers. As a result, it has been said, "Superb sermons were often thundered from pulpits on the importance of raising good tobacco and the moral necessity of curing it properly." Also important, as far as the clergy were concerned, was to avoid taking a post in a church that sat on rocky or acidic or nutrient-starved soil, or that stood adjacent to a poorly managed plantation. "Some parishes," explained the Reverend Hugh Jones of such priorities, "are long vacant upon account of the badness of the Tobacco."

A preacher of the late seventeenth century had several ways to augment his income. He could perform a funeral for four hundred pounds of leaf or a wedding for two hundred. For less important ceremonies, or duties of some other nature that seemed to require a divine intermediary, he could negotiate a smaller fee with the interested party.

But this was moonlighting. A clergyman's main source of income was his salary, and by the 1750s it had become fixed in Virginia at 16,000 pounds of tobacco per year. Most parishioners, the people whose tithes provided the leaf, thought this was too much; most men of the cloth, not having renounced earthly goods to the extent of their brethren in monastic life, thought it too little. The historian Thomas Jefferson Wertenbaker agrees with the latter:

> All in all the livings of the Virginia clergy were most inadequate. It was a common complaint among them that they could not attend properly to their duties because they were harassed by poverty. They could not supply themselves with books. Many of them were compelled to remain single, for they could not afford to support a family.

But there did not seem to be any means of redress for them. In most cases, the clergy of colonial times saw no choice other than to keep their grumblings to a minimum and make sure they did nothing to anger churchmen, who might try to get even by reducing their pay openly, reducing it surreptitiously by shortweighting them, or, in extreme cases, refusing to hand over any leaf at all. It was, for God's agents in Virginia, a dispiriting predicament.

Then in 1755, Virginia suffered through one of its worst droughts ever. Weeks went by without so much as a drop of rain, months without any significant accumulation, and through it all the sun beat down with ferocious intensity. There were few breezes, and when they stirred it was to blow the parched topsoil of one farm onto the next, like a shower of ashes from one old fire to another. Thousands of acres of normally fertile land were either destroyed or damaged so much that the crops struggling to grow out of them were at best hopelessly stunted, at worst stillborn.

As a result, the planters' income also dried up. Some parishes made a sincere effort to find the leaf to pay their preachers; others did not, simply notifying them that there was not enough tobacco to go around this year and they would have to take whatever could be spared, if anything. And that got the planters to thinking about future years and the possibility—even the likelihood—of more shortfalls. Something had to be done, they decided, to spread the risks of agriculture among the clergy, whose guaranteed income meant that they felt none of the harmful effects of a time of blight.

That something happened in 1758. The Virginia Assembly passed the Two-Penny Act, which allowed all debts in tobacco to be paid off at the modest rate of twopence per pound of leaf. The clerics, caught by surprise, conceded that the amount was a fair one in a bad year. But most years were good, they maintained; in fact, tobacco production had risen substantially since the drought of '55, and there was every reason to believe it would continue to rise, reaching new highs in the future and bringing even more prosperity to those who grew it. Why, the preachers asked, should they be penalized for an event from which the colony had already recovered? Why should salvation, than which there was no more prized commodity among Americans, suddenly be provided at a discount?

The drought was the stimulus for the Two-Penny Act; it was not, however, the only reason. Hard feelings had existed for a long time in Virginia between men of the soil and men of the cloth, in some cases going back as far as the very beginning of their relationship in the New World. Many of the farmers and planters believed that the clergymen thought of themselves as a privileged class, which was apparent not only in their demand for 16,000 pounds of leaf a year, come what may, but in the paltry sums they contributed out of that pay for defense against the French and Indians.

Further, and perhaps more important, there was prejudice at work here. The more successful tobacco farmers, having become an elite class themselves—at least, economically—believed that the clergy were their social inferiors, and no one can abide a social inferior putting on airs and demanding pay raises. Even worse, the farmers had come to America at least in part to escape from the Anglican faith that some of the clergy espoused, while others, not so hostile toward the faith at the start, had now begun to think of it as not particularly relevant to life in the wilds of the New World. It was a religion for a monarchy, not a budding republic, and was perhaps even a means by which one monarch in particular sought to control them, the church being an extension of an unjust colonial government originating with the king of England.

The clergy, aware of all these factors, appealed to the Assembly for relief from the Two-Penny Act. It did not provide any. Then they appealed to the Motherland, and here they received a warmer reception. It seems the Privy Council had ruled some years earlier that no revenue measure could be passed in the colonies unless the council first gave its assent. It had not done so in this case and therefore was pleased to act on the clergy's complaint, hoping to teach the insubordinate Virginia planters a lesson. The council did not overthrow the Two-Penny Act, which was due to expire after twelve months anyhow, but it made certain that the Assembly did not extend it and went on to rule that the colony's preachers were entitled to back pay for the full year of the act's existence.

By now, the Virginians were used to royal decisions going against them. This did not, however, make the reversals easier to take. Among those who took this one especially hard was a young man from Hanover County who had been born a rustic and still like to play the part on occasion, exaggerating his backwoods drawl among the front-woods types who were his natural constituency. True, he had had no formal schooling, and, true, when he tried his hand at business he went bankrupt in a remarkably short time, perhaps because he spent too much of the workday talking to customers, too little with the balance sheets. But the reason was that he "seemed beguiled by the thrust and parry of any argument and determined not to let his job get in the way of a good debate." He was not an incompetent young man but, rather, a bright one in search of the right opportunity.

He dared to think he had found it in his most recent fling at vocation, as an attorney. He had, after all, "a smoldering eye, a querulous voice, and a marvelous gift of the gab." His name was Patrick Henry, and when the Privy Council came down on the clergy's side on the Two-Penny Act, he argued, not for the last time, that the crown had no business meddling in America's affairs. He had not, however, established much of a reputation at the time.

Some of the clergy, afraid of a backlash by the members of their flock, decided not to seek the back pay that the act granted them. Others had no such inhibition. They were not greedy, simply desirous of all to which they believed they were entitled by law or custom. The Privy Council having defined justice in their favor, they were determined that it be enacted. They went to court. They filed lawsuits. These events, and the grievances that led up to them, are known to history as the Parson's Cause.

The third of the suits to come to trial was heard in Henry's Hanover County, about eighteen miles north of Richmond, in 1763. The plaintiff, an Anglican named James Maury, has been described as "a dignified gentleman of Huguenot descent." Beyond that, little is known of the man other than that, believing firmly in the rightness of his cause, he expected to win. And technically he did, but only technically, perhaps because his case was tried in the most unsympathetic of all possible venues.

For the preceding decade, if not longer, Hanover County had been one of the leading centers of the Great Awakening, a series of religious revivals that, according to the historian Paul Johnson, "seems to have begun among the German immigrants, reflecting a spirit of thankfulness for their delivery from European poverty and their happy coming into the Promised Land."

Other immigrant groups were also thankful; like the Germans, they were escaping from despotic rulers, secular or ecclesiastical or both. As a result, a patriotism-soaked religious fervor had spread rapidly through the colonies, with a depth of commitment and raucousness of expression that were distinctly American. The Great Awakening emphasized fiery preaching and lay participation in services; it meant for worship to be an informal thing, unbound by either pomp or bureaucracy. It also stressed the importance of individual spiritual experience over church doctrine, especially doctrine of the stodgy old Anglican church. Hanover Countians had not much cared for that particular faith *before* the Great Awakening came along to roil their juices; these days, they could be positively hostile. Obviously, this did not bode well for poor Reverend Maury.

Neither did the composition of the jury, most of whom seemed to be chosen on the basis of their resentment of either the Anglican church or the rates of compensation for its ministers or both.

Nonetheless, to the amazement of virtually all, judge and jury found in Maury's favor, and a second trial was scheduled to determine how much money he would receive. The tithe-paying members of his congregation were more than disappointed by the verdict; they were fearful of a large settlement, to which they objected both in principle and on the grounds that, still reeling from the effects of the drought of '55, they could not afford it. They were further disheartened when their attorney, John Lewis, resigned. His clients begged him to continue through the second trial, to try to change the judge's mind. He refused. Great Awakening or not, he thought the case unwinnable and wanted nothing more to do with it.

The parishioners needed to replace Lewis, and they needed to do it fast. They did not have time to quibble about qualifications or to insist on a congenial spirit. To make matters even more difficult for them, they determined that even after pooling their resources, they could not afford more than fifteen shillings. Who can we get for so meager an amount? they wondered. Maybe they should give the fifteen shillings to Maury—just hand it over, then go home, plant a few more acres of weed, and hope for the best.

One man was available to them, a single attorney in the whole of Hanover County. Patrick Henry did not care about the money; he cared about the people who would listen to him in the courtroom, the attention that would be paid. In many ways, he was a surprising choice to challenge the Privy Council's ruling, and not just because of his unimpressive background. He also happened to be the son of John Henry, a devout Anglican, "one of the few men around Hanover who had been to college"—and the presiding judge at the first Maury trial. It was Henry's dad, in other words, who had found in the reverend's favor and so discouraged John Lewis.

But the defendants did not have time to fret about that, either, and Patrick Henry was ushered onto the center stage of colonial affairs, determined to make the best of his debut.

According to contemporary accounts, Henry's summation to the second jury started slowly. His phrasing was awkward, his voice low and indistinct; many in the courtroom could not hear him, and those who could did not seem to be impressed. You get what you pay for, they might have thought, and Henry was a fifteen-shilling barrister.

But he rose quickly in value, gaining confidence as he went along, speaking more boldly, standing more erect, presenting himself in a more assertive manner, and thereby compelling the notice of all within earshot. Soon, says Henry's biographer Richard R. Beeman, the jurors were

> taken captive, and [were] so delighted with their captivity, that they followed implicitly, withersoever he led them; that at his bidding, their tears flowed from pity, and their cheeks flushed with indignation, that when it was over, they felt as if they had been awaked from some ecstatic dream, of which they were unable to recall or connect the particulars.

One suspects Beeman of hyperbole. Maybe he got the idea from his subject, who, while addressing the jurors, soared into realms of verbal excess himself, although with the intent to denigrate, not glorify. Maury and his ilk were not preachers of the Holy Word, Henry told the jurors, but "rapacious harpies," men who "would, were their powers equal to their will, snatch from the hearth of their honest parishioner his last hoe-cake, from the widow and her orphan children their last milch cow! The last bed, nay, the last blanket from the lying-in woman!"

Even by the generous standards of lawyerly closing arguments, Henry was overdoing it, coming closer to slander than summary. But he was well aware of the fact and did so deliberately, with an eye to the future. He actually admitted as much to Maury, taking him aside after the trial and telling him that he had been more concerned with making an impression than with approximating the truth. He was, in other words, putting on a show, promoting himself no less than trying the case: a twenty-first–century attitude in an eighteenth-century courtroom.

Maury, for whom the case was a more serious matter than that, raged at Henry's admission. Henry was an opportunist, he said, a ruffian, a man who "thinks the ready road to popularity is to trample on religion and on the prerogatives of the king."

But it worked. Henry is said to have picked up more than 160 new clients within a year of the Parson's Cause because of all the publicity. More important in the long run, he became a figure in Virginia and elsewhere, a spokesman for colonial grievance and a fixture in the history books of later generations.

But we are getting ahead of our story. In his remarks to the jury, Henry did not merely trample on the prerogatives of George III; he savaged the king's name. George was no longer "the father of his people," Henry declaimed. Now he "degenerates into a tyrant and forfeits all rights to his subjects' obedience." It was incendiary language, breathtaking in its audacity, a prelude to combat.

Patrick Henry spoke to the jury in Hanover County for an hour. When he finished, the courthouse was hushed. His arguments inspired some, embarrassed others, and made an impact on all. Maury's attorney then offered his own concluding comments, although for less time and to considerably less effect. The judge gave his instructions, going over the specifics carefully, and the jurors retired to deliberate.

They were back in minutes. They had barely needed to sit down, much less talk among themselves. They were bound by the law to find for the plaintiff, but their verdict would reveal their loathing for the Privy Council's meddling in American affairs. They awarded to Reverend James Maury back wages in the total amount of one penny!

The defendants, those suddenly relieved churchgoers certain that, hard times or not, they could more than likely scrape up a single cent among them, sprang from their seats, letting out whoops of gratitude and relief and jostling one another in celebration. They "lifted Patrick Henry, as a hero is lifted, and carried him in wild excitement on their shoulders."

Henry's words would incite similar passions in the years to come.

BY THIS TIME, tobacco had transcended its genre yet again. More than a crop *and* a currency, it was now a way of life, a society unto itself for those who grew it as well as those who smoked it and smelled it in the air around them—and this included almost every human being in every British colony in every single settled nook of the New World. The leaf was income, recreation, relaxation. It was a tonic at the start of the day and a tool of reflection at the end. Tobacco shaped a person's ideas and attitudes; it influenced his choice of friends and his style of speaking and the allocation of his time. As the first successful American export, as well as a prized personal possession, it gave a clearer purpose to the seasons than they had ever had before. This was not corn, not wheat, not barley, not anything so common. This was tobacco, brown gold, and the weeks of planting and harvesting and curing became the events around which all others in the community revolved. At this stage in America's growth, it was the tobacco plant, not the bald eagle, that would have been the more fitting symbol of the colonial experience.

And the plantation system supporting the weed was a testing ground for the eventual United States in two vital and very different ways. First, it tutored a number of the men who made the revolution. Few of them came from families with impressive backgrounds or a tradition of learning; leadership was not in the genes. Rather, it was forced upon them by vocation, and most proved up to the challenge. Owning and running a plantation required executive skills of a high order, skills that were also helpful, it not mandatory, in founding and maintaining an independent nation. Acquiring the skills, these men were also acquiring a degree of self-regard they would not have known otherwise, which is to say that, in addition to their newfound knowledge and abilities, they developed the confidence to use those abilities under pressure. "No amount of book-learning," writes Sydney George Fisher, "no college curriculum imitated from plodding, mystical Germans, no cramming or examinations, and no system of gymnastic exercises can be even a substitute for that Virginia life which inspired with vigor, freshness, and creative power the great men who formed the Union and the Constitution."

The slaves who worked on the plantations lived under a different set of pressures, however, and in this way, too, was the tobacco culture a testing ground for American democracy. The owners succeeded at leadership; they failed at decency. They allowed forced labor to become the foundation of their earnings, but the very presence of slaves made a mockery of the principles they so publicly professed. Slaves accused America of hypocrisy without opening their mouths. They charged it with dishonesty by simply playing their parts. In truth, the new country was guilty of even worse.

The plantation owner did not merely abuse his slaves in body; he broke their spirits. It was part of the strategy. The owner believed that men and women who have been stripped of their self-worth are men and women who will not try to escape from drudgery, even painful and debasing drudgery, or rebel against those who impose it. And so the planter took away the slaves' spouses and split up their children and segregated them from their friends and denied them contact with their homelands and in many cases provided them with housing that was barely habitable and food that could barely be eaten.

If the slaves complained, or transgressed in some other manner, they were broken even further:

> Even as planters employed the rod, the lash, the branding iron, and the fist with increased regularity, they invented new punishments that would humiliate and demoralize as well as correct. What else can one make of William Byrd's forcing a slave bedwetter to drink "a pint of piss" or Joseph Ball's placement of a metal bit in the mouth of persistent runaways.

The slaves cried out at this barbarity, this violation not only of all that was compassionate but of all that was sensible in human behavior. In time, their cries became crises of conscience that continue in the United States, although in less tragic circumstances, to the present day.

THE FIRST PLANTATIONS were built on riverbanks, each with its own dock so that traders from England could sail right up to the estate and deal directly with a grower rather than a middleman. Tobacco and money changed hands without delay. Then the traders scheduled their next visit and weighed anchor. They sailed downriver to another plantation, a few miles away, for an equally efficient transaction.

But before long, the riverfront property was gone, and the tobacco men had to be more resourceful. They erected their plantations inland, sometimes many miles into the forest or swamp, and had their slaves cut roads through the trees and wild shrubbery to get their crops to water. It was not easy. First the slaves packed the tobacco into barrels called hogsheads, each about thirty inches in diameter by forty-eight inches in length, each capable of holding more than 1,000 pounds of leaf, and each with an axle attacked to the back end. Then horses were hitched to the front, turning the hogsheads into small, temporary wagons. The horses towed the hogsheads down the roads, making a clatter as they churned up clouds of dust and frightened the small animals. At harvest time, or when traders were waiting at the docks, the dust never settled, the animals never relaxed.

Prior to being loaded onto ships for the Atlantic passage, the hogsheads were often pried open and inspected. If the tobacco was "sound, well conditioned, merchantable, and clear of trash," it went on its way. If it was not, it was returned to the plantation, where it would be discarded, and relations between planter and trader would deteriorate. If the tobacco was not in pristine condition and the inspector, having taken a bribe from the planter, approved it anyhow, he ran the risk being put to death "without benefit of clergy," which meant not only that his days on earth were cursed, but that he would spend all the days of eternity burning in hell—a stiff price to pay for a few extra bucks. After a while, the punishment was reduced to a jail term of between one and ten years, depending on the amount of trash the tobacco contained. Still, the message was clear: The weed matters; disturb its purity at your own peril.

Plantations were not, as they are sometimes thought of today, like farms, even big farms. They were more elaborate than that, more complex to organize and operate, and more varied in function. On occasion, foreign visitors would drop in on a Virginia plantation and relate to friends at home that it was like visiting a small town, and in many ways, it was. It makes more sense, though, to think of a plantation as a feudal domain of some sort, as

> the counterpart of the medieval manor. The planter was the repository of social dignity, of judicial power, of political leadership for his neighborhood,

> in quite the same fashion as the eighteenth-century English squire whom he strove to emulate. Both were survivals of the feudal tradition. The presence of the Negro slaves, however, gave an added intensity to feudal feeling in the South. Here was a factor far closer to the serf of the Middle Ages than anything in Georgian England.

The centerpiece of the plantation, if the owner could afford it, was a palatial residence for his family and himself. But, perhaps being aware of the feudal comparisons and not wanting to push them too far in a land that was already boasting of egalitarian ideals, he settled for referring to his home not as a palace or a manor but simply as the big house, a term of grandiosity through understatement. Writes historian Catherine Clinton: "Dining rooms, parlors, libraries, music rooms, and sitting rooms were located off the central hall; bedrooms and nurseries, on the upper floors. Many mansions had porches on both upper and lower stories. Most homes were equipped with fireplaces in every room. A handsome staircase in the center of the house was often supplemented by a back stairs for servant use."

In addition to this magnificent abode, the plantation had a smokehouse, a slaughterhouse, a detached kitchen, barns, stables, workshops, curing and storages areas for the tobacco, perhaps a chapel, and, set apart from the other structures, the cabins for slaves. Also populating the grounds were "carpenters, coopers, sawyers, blacksmiths, tanners, curriers, shoemakers, spinners, weavers and distillers," almost all of whom were white men who could come and go as they pleased.

Not all of the land was given over to tobacco. There were gardens for vegetables, orchards for the apples and peaches that were made into brandy as well as eaten whole, and pastures where animals fattened themselves for the platter. The plantation not only raised and grew and prepared its own food, it made its own clothes, built and repaired its own buildings and machinery, and provided its own entertainment, such as sing-alongs, dances, horse races, and the occasional, spur-of-the-moment athletic event, usually involving white people sitting back comfortably and sipping on a beverage while they bet on the performance of blacks. It was possible to leave the plantation for business or pleasure, sometimes even desirable. But seldom was it necessary.

The owner was in charge of all matters relating to business. He delegated many of the day-to-day responsibilities to an overseer, but it was on the owner's shoulders that the fate of the plantation ultimately rested, and so it was he who approved decisions relating to expenses and hiring, he who dealt with local government and foreign traders, he who made sense of the account books and, if he had the time, sometimes experimented with new methods of cultivating or curing the tobacco, always striving for a better product, a better price. The "larger and more successful Virginia planters of the 18th century," Daniel Boorstin tells us, "were interested in natural history, had a respectable knowledge of medical remedies and mechanics, were at home in meteorology, and felt obliged to know the law."

The smaller planters were no fools, either, even though they might farm but a few hundred acres, possibly even less. True plantations, though, like the estates of the feudal seigneurs, sprawled over thousands of acres, some of which would be lying fallow in a given season and some of which would be growing crops other than tobacco, this acreage also taking a break for a year or two from the leaf's soil-enervating demands. When a field reached the point at which the breaks did not revitalize it anymore, it was retired, literally put out to pasture or maybe even allowed to go wild again. The slaves would then be ordered to hack a new field out of the forest to replace it—tobacco punishing land, punishing labor, indiscriminate in its exactions.

The largest plantation of the colonial era belonged to Robert Carter, known to one and all not as mere lord but by a higher designation. By the time of his death in 1733, "King" Carter had accumulated not only 330,000 acres of choice Virginia soil, 2,000 horses, and seven hundred slaves, but more power than any other man in the colony, perhaps in all the colonies put together. An imperious sort, Carter held a variety of political offices, almost any that he wanted at any time he wanted it: justice of the peace, member of the House of Burgesses, Speaker of the House, and commander of the local militia unit, among others. So highly regarded was he by his fellow colonists, or so intimidated were they by him, that Sunday services at his church did not begin, nor did the clergyman even take the altar, until the King and his family had seated themselves.

When he was not being self-important, he was being gregarious. James MacGregor Burns says that

> Carter and his friends thoroughly enjoyed the rich offerings of Virginia's rural life—hunting, racing, fishing, riding, drinking, gambling, cockfighting. But Carter's Nomini Hall [the big house] overflowed with the sounds of learned discussions and lively music, of polite socializing and stately dancing.

Another esteemed planter was William Byrd II, also known as William Byrd of Westover, also known as the black bedwetter's worst nightmare. Possessor of some 179,000 acres, he was almost as influential a figure as Carter, but not quite so overbearing—not, at least, to those of the same skin color.

"The Merchants of England," he once said, speaking of his fellow plantation moguls, "take care that none of us grow very rich." Then he added that "the felicity of the Clymate hinders us from being very poor."

Byrd was one of early America's more learned citizens, and it was for this reason that his return to the colonies in 1726 after a decade abroad was a painful one. As Bernard Bailyn points out, "There was no one to respond to his wit, his satire; no one to acknowledge his intellectual achievements, no way to establish his worth as a man of letters, as a man of the world. He was no longer in the world. Nostalgically, he kept his rooms in London, practiced his languages—every day some Greek and Latin and a bit of Hebrew—read diligently, remorselessly, in several Europeans languages, built up his library into a formidable collection of over three thousand titles." It was said to be the second largest in the colonies. Only the Reverend Cotton Mather's, in Massachusetts, exceeded it.

A writer of some skill himself, as well as being a knowledgeable amateur physician, Byrd combined the two talents in his pamphlet called *A Discourse concerning the Plague with Some Preservatives against It*, the author identifying himself as "A Lover of Mankind." Like many of his British forebears, Byrd swore that tobacco was the ultimate "preservative," writing that it was responsible for the plague's not having made an appearance in England for more than half a century.

And it was not just the plague. Byrd believed that other ailments would yield to the leaf's restorative powers, and his fellow Americans agreed, both in his generation and those to follow. Late in the eighteenth century, when epidemics of yellow fever struck Philadelphia twice within five years, the universal treatment was cigar smoke, with some citizens relying on it to such an extent that tobacco sellers reported doubling and even tripling their business.

But merely smoking the weed, Byrd insisted, was not enough. Sensible human beings should go further:

> We shou'd wear it about our clothes, and about our coaches. We shou'd hang bundles of it round our beds, and in the apartments where we most converse. If we have an aversion to smoking, it would be prudent to burn some leaves of tobacco in our dining room, lest we swallow infection in our meat.

Other plantation families bore names that were famous at the time and have remained so, to one degree or another, ever since: Fitzhugh, Lee, Randolph, Harrison, and Nelson, to cite a few. Yet perhaps the most admired grower of the pre–Revolutionary War period was a man whose name has not withstood the years, a man whose plantation was relatively small yet whose output was the talk of planters not only throughout the colonies but abroad. His name was Edward

Digges, and he turned out "the lightest, mildest leaf grown in Virginia." Because the hogsheads in which Digges shipped his product were stamped with his initials, his tobacco was known far and wide as "E Dees," and both during his lifetime and afterward it brought higher prices from English traders and more praise from English smokers than the tobacco of any other American grower.

According to Robert K. Heimann, these men and others like them "were to Virginia what the cattle barons would be to the early West and the oil millionaires to Texas." Before the nation even was a nation, they were its first aristocracy, although bloodlines had nothing to do with it. They became able men by running their businesses. They became self-respecting men by running their businesses well. There was no denying their ability as a class, the extent to which they were prepared for the tumultuous events to come. "Trained in the management of self-sufficient plantations supporting up to 1,000 souls apiece," Heimann goes on, "they passed easily into the management of whole colonies and the conduct of international affairs—first under, then against, the haughty kings of England."

OF COURSE, THE AMERICANS were haughty, too, trafficking in slaves as they did, so many of them so casual about it, certain that the practice could withstand scrutiny, even if they had no intention of providing the scrutiny themselves. Some did not believe that blacks were quite human. Others assumed, as custom dictated and the law affirmed, that slaves were property, just as a chair or an anvil or a hogshead was property, and regarded the people as unfeelingly as they did the inanimate objects.

Still others, despite their reliance on slavery, were troubled by it, and some—notably, Thomas Jefferson—turned their angst into a virtual school of philosophy. "You know," he wrote to a friend in France in 1788, "that nobody wishes more ardently to see an abolition not only of the trade but of the conditions of slavery." More than once, Jefferson proposed that slaves be freed at birth, educated and prepared for careers at public expense, and then dispatched to live among their own kind somewhere west of the Mississippi, or possibly in a foreign land such as Santo Domingo or Africa. Yet he was a slaveowner all his life, in all likelihood the father of several children by a slave, and he admitted to various companions at various times that abolition might create more problems than it solved.

One of those problems would have to do with tobacco. The historian Stephen Ambrose is right when he says that "slavery was critical to tobacco planters because their agricultural practices were so wasteful and labor intensive": wasteful, as has been demonstrated, because the weed so prodigiously used up the land; labor-intensive because the crop needed tending from winter to fall. It was an axiom among growers that tobacco "requires not skilled hands but many hands."

The first thing the hands did was plant the tobacco in seedbeds in January or February. Then, after replanting a time or two, they transferred the crop to the fields in April or May, even though "the loss of seedlings from cold, disease, and insects was such that ten times more seeds were planted than the fields would have room for."

Situating the plants in the ground was a process both time-consuming and arduous. The hands placed each plant about two or three feet from its neighbor; any farther apart would have been a waste of space, any closer undue crowding. As the plants matured, the hands worked the ground around them as meticulously as sculptors working their clay: hoeing, weeding, priming, topping (cutting the tops off the plants so they would not flower), hilling and re-hilling, all the while checking for worms and other insects and making sure the maximum number of leaves per plant did not pass a dozen. If it did, the excess leaves were snipped off; a single plant could not support too much growth. In addition to this vigilance, a good tobacco crop demanded just the right amount of moisture and just the right amount of sunlight—in other words, just the right amount of good fortune from above.

The harvest came in late summer. The leaves were picked and cleaned, then dried and cured, usually by exposing them to air or low-burning fires under carefully controlled conditions in structures built especially for this purpose. Sometimes the curing took four weeks, sometimes as long as eight; daily monitoring was necessary to regulate the air flow or keep the fires at the precise temperature and height. Finally, the tobacco was packed into hogsheads and either stored for later use on the plantation or rolled down the road to the river. It was reckoned that a plantation needed one slave for every three acres it devoted to the weed, a much higher proportion of human to field than was required by any other agricultural commodity.

Initially, though, the work was done not by slaves but by indentured servants, white men and sometimes women, who had been carted to the New World from England, agreeing to be pieces of property themselves for a time because they were too poor to travel to America in more favorable circumstances. Some of them were criminals, paying their debt to society. Then there were those who, although few in number, had been forced into an indentured state, kidnapped by merchants from their homes or shops or fields, stolen in one place to be fenced in another.

For a while, these laborers were a going concern in the New World. They were also a perishable one, as in the seventeenth century "it was not uncommon for one-third of the servants to perish before the ship arrived at an American port," the result of massive overcrowding, poor nutrition, and savage treatment at the hands of merchants.

The problem for the plantation owners was that, regardless of how the servants were procured, there were never enough of them to go around. Further, the terms of their bondage usually required that they be set free after a period of between four and seven years. Or they might escape and with their white skins blend easily into the populace. It did not take long for employers to decide that they were better off with cheap, abundant blacks whose services they could secure for a lifetime and whose conspicuousness made them easily retrievable.

As a result, the vile trade boomed. From 1710 to 1769, more than 53,000 slaves were imported by tobacco growers in Virginia alone. In 1715, the total population of the colony was 95,000; almost a quarter of them were blacks, and virtually all the blacks were slaves. By 1753, the population had risen to 168,000, of whom close to 40 percent were dark-skinned. It was too much even for William Byrd, who complained that "they import so many Negroes hither, that I fear this colony will some time or other be confirmed by the name of New Guinea." So great, in fact, was the increase in the slave population that it led to a new breed of abolitionist in Virginia: the white man opposed to slavery not for religious or humanitarian reasons but for fear of what might happen to him and his fellow whites if the slaves one day came to outnumber them.

Working from figures like these, a few historians have gone further than Ambrose and concluded that because tobacco needed slavery, it was the primary reason for slavery's presence in North America. If there had been no weed, the argument goes, there would have been no slaves, or far fewer, at least in the long run. The premise is simple, obvious, and wrong.

It is true that plantation owners would not have been able to operate on so large a scale without slaves, and it is equally true that the larger the operation, the more entrenched slavery became. But the first slaves in the Western Hemisphere did not pick tobacco. It was sugar cane that brought them from Africa to this part of the world, and it was the plantations of the Caribbean on which they toiled, places where the seasons never changed and the workday never ended.

The slaves buried cane cuttings and chopped down the mature stalks. They crushed the stalks and sprayed them and shredded them. They hauled away the fibers, then boiled the cane juice in huge vats on hundred-degree days, their skin feeling as if it, too, were afire. Then they refined and processed some more, rinsing off molasses film and dissolving sugar crystals and filtering the remaining solution, and after that they packed and stacked and readied for export. And they did it again and again, day after day, season after season, death after death. "The mortality rate of slaves hacking away under a pitiless tropical sun was simply staggering," writes Ron Chernow. "Three out of five died within five years of arrival, and slave owners needed to replenish their fields constantly with new victims."

And it all happened within a short sea trip from America, the proximity ensuring that the colonists would learn of the practice, in some cases witness the practice, and find an excuse to import slaves themselves, whether tobacco required them or not.

And, in fact, it was other tasks that awaited at the outset. Slaves started out in the colonies as artisans and shoemakers, tenders of livestock, and domestic servants. In some villages, they acted as the equivalent of handymen; in others, they assisted shopkeepers, although they did not as a rule wait on customers. Some of the women were midwives; a few of the men did heavy lifting on the docks. And whereas in time slaves were delivered to Virginia and Maryland largely for tobacco, they were also being assigned to other commodities. "According to the 1810 census," writes Mark Kurlansky, "Kanawha county [Virginia] had 352 slaves, but by 1850, 3,140 slaves lived in the county, mostly assigned to saltworks." And slaves were being dropped off in South Carolina and Georgia to help with the rice and indigo crops.

Furthermore, some of the first slaves in the colonies were not imported blacks but the members of native tribes. "In New England," we learn from Arthur Schlesinger Sr., "they were used chiefly as domestics, working as a matter of course even for such notables in religious and public life as Governor John Winthrop, Roger Williams, and Increase and Cotton Mather." To the south, the natives hunted and fished and turned their yield over to their masters.

Tobacco did not, then, cause slavery in the New World. It did not guarantee slavery's ongoing presence and economic importance. It did not worsen the conditions of slaves and thereby lead to an ignoble institution's becoming even more shameful. What it did do was make slavery more widespread than it would otherwise have been, but only until the Revolutionary War, at which point the number of slaves in the tobacco fields exceeded the number in all other venues put together, and production of the weed was at a peak.

But after the war, the economy of the southern United States began a long, slow, and crucial transformation, at the end of which tobacco no longer ruled. It was not that the leaf became less popular—to the contrary. It would continue to attract customers for another century and a half. It was not that the labor pool dried up; in fact, more slaves were deposited on American shores after the war than before. And it was not that too many of the fields were depleted and could not stand the strain of tobacco anymore, although that was part of it. The rest had to do with technology, finance, and culture, which combined in their own inadvertent way to anoint a new king of the Southern soil. This ruler was called cotton, and its reign was not only more lasting than tobacco's but more important to the promotion of slavery.

Like the weed, cotton depended on unskilled hands, the more the better. This was true when the seeds had to be separated from the fibers manually, and it remained true after Eli Whitney invented the cotton gin, a marvel of a machine that utilized rapidly spinning blades with teeth on them to do the chore. But someone still had to plant the cotton. Someone still had to pick it. Someone had to put it in the gin and get the gin running. Someone had to remove it and prepare it for shipping.

Before Whitney came along, though, no one shipped very much. Cotton was not profitable—not as much as many other crops, at least. Too many hands were needed, and it took them too long to pluck out the seeds, dispose of them, and bag the fibers. Whitney believed his device would change all that. He claimed that "one man will clean ten times as much cotton as ... any other way before known and ... clean it much better. ... This machine can be turned by water ... or a horse, with the greatest ease, and one man and a horse will do more than fifty men with the old [method]." Eli Whitney, to the surprise of no one who knew him, was bragging. He was also right.

But at the same time that the owners of cotton plantations were equipping their men with gins in the South, new textile machinery was being invented and installed in factories in the North. This was the other revolution of the era—the industrial one that would make America wealthy as opposed to the military revolution that made it free—and things were at their most revolutionary in the textile business. The new machines were even more marvelous than the cotton gin, agents of the future set down in the present. They enabled fabrics to be manufactured more rapidly and cheaply, and of higher and more uniform quality, than ever before. This, in turn, whetted demand, which at first meant more labor for the men using Whitney's labor-saving device and then meant more men to use it.

And with President Jefferson's purchase of Louisiana from the French in 1803, almost 828,000 square miles were added to the United States, ranging from the Gulf of Mexico to Canada, from the Mississippi River to the Pacific Ocean. More of the land went to the new king than to the old, and those who grew the cotton in the expanded territory bought slaves not only from abroad but from tobacco plantations, which, for reasons soon to be explained, did not need as many anymore. In fact, by the 1830s, cotton employed so many blacks that Josiah Quincy, the president of Harvard College, attributed to it the "dangerous and rising tyranny" that slavery had become.

He exaggerated. Like many others at the time who were enraged by slavery and trying to find a solution, he looked only at the surface of events. In the final analysis, cotton cannot be blamed for the appalling state of black bondage any more than tobacco can. Agriculture cannot be blamed. Nonfarm employment cannot be blamed. Slavery was not the fault of marketplace exigencies or conditions of geography or the failures of a particular people at a particular time in their history.

It is, rather, human nature that must answer for slavery, the tendency of those who may otherwise be decent men and women to behave, in certain settings, with gross disregard for the humanity of others, especially when those others are a different sort from themselves. For slavery is injustice, plain and simple and unconscionable, and injustice has been the curse of all peoples at one time or another, to one degree or another, and will continue to be until the last tobacco field is plowed under and the last pipe or cigar or cigarette is extinguished. Injustice is far easier to record than it is to eradicate. It is far easier to rue than it is to defend.

But this particular injustice is not, in the final analysis, tobacco's fault.

AS WAR BETWEEN America and England grew closer, the colonies grew closer to one another. Earlier, they had been like siblings in a family that gave lip service to their ties but were not especially close. Sometimes they bickered, went off in huffs, turned unresponsive. Often one colony just ignored its neighbor, going about its own business and having no particular desire to defend or even explain itself.

Then the crisis came, and the colonies were forced not only to recognize their common cause but to take pride in it. They began to depend on one another and to communicate more than they ever had in the past, organizing themselves into a kind of unofficial union through Committees of Correspondence, which met on regular bases to exchange information and keep up morale. These were the first American support groups.

December 17, 1773. Some citizens of Boston gather in the back room of a newspaper office and darken their faces with burned cork, paint, or grime from the floor of a local blacksmith's shop. Then they dress up like Mohawk Indians. All the while, in the words of Samuel Adams about a similar occasion, they probably "smoke tobacco till you cannot see from one end of the garret to the other." Some of them stay in the office; others gather in their favorite taverns to drink rum until nightfall. They might be getting up the nerve for their mission; they might also be blotting out awareness of it, feeling guilty about the drastic deed they are about to perform. They are, after all, still Englishmen.

When the moment seems right, the ersatz Indians assemble on the wharf, where they climb aboard three British ships that are anchored in the harbor. On the ships are a total of 340 chests. The colonists carry each of them to the rails and dump the contents, thousands and thousands of pounds of tea, into the water. It is a protest against both the tax on the product and the monopoly that British traders enjoy on its sale. "The next day," writes A. J. Langguth, "Boston's harbor looked less like a seaport and more like a vast dank beach. Shaped into dunes, the tea lay upon the water and clogged the sea lanes. Sailors had to row out to churn the sodden heaps and push them farther out to sea."

The crown was startled, infuriated, virtually uncomprehending, with the king himself denouncing "the violent and outrageous proceedings" that would henceforth be known as the Boston Tea Party. Parliament closed the port of Boston, declared the colonial charter no longer valid, and replaced a number of popularly elected officials with a military governor. It was, in modern terms, the imposition of martial law.

But it did not work. Parliament's action was, if anything, fuel on dissent's already raging flames. In Virginia, the Committee of Correspondence no sooner learned of Boston's punishment than it called for a day of fasting and prayer, "devoutly to implore the Divine interposition for averting the heavy calamity which threatens destruction to our civil rights, and the evils of civil war." However, the crown thought no more highly of Virginia's compassion than it did of Boston's midnight raid. Two days later, the colony's governor, Lord Dunmore, told the Assembly that it was no longer in business.

When members of the Committee of Correspondence, many of whom also held Assembly seats, next gathered, it was in full realization that a formal outbreak of hostilities could not be far away. To some, this meant they should proceed with caution. But to others, it meant retaliation, hastening the inevitable, and these men seemed to be the majority. They voted to stop all imports from England, even though this would also have dire consequences domestically, and they urged the Committees of Correspondence in other colonies to agree to a Continental Congress in Philadelphia in 1774. Among other things, the Congress would consider a declaration of independence.

It was a decisive step, drastic even, and the Virginians knew it. For this reason, they took it in their usual meeting place, the chamber in which they so often sat to discuss matters both solemn and social, where they felt both comfortable and stimulated, no matter what the occasion: the Sir Walter Raleigh

Tavern in Richmond. Alistair Cooke, well aware of the tavern's standing, calls it "the Old South Meeting House of the Old South."

The tavern was an aptly named site to formalize rebellion, for tobacco had remained a point of dispute between the colonies and the Motherland ever since the days of John Rolfe and his first successes with homegrown *tabacum*. Charles I, who followed James I to the throne, was almost as bitter a foe of tobacco as his predecessor. Twelve years into his reign he told the Privy Council that Virginia was "wholly built upon smoke," and he was much displeased about it. The council passed the word along to the Americans, who, of course, had heard it before and paid no more attention to it this time than they had previously.

Later in the seventeenth century, Virginia planters found themselves the victims of overproduction. Too much crop, too little profit. The Assembly decided to act, announcing that it would consider a resolution to restrict yields and thereby boost earnings. Its members were leery of tampering with the marketplace, which even then was thought by some to operate with its own internal logic, resistant to outside forces. But many of the assemblymen were also farmers and as such had become desperate. Practical considerations overwhelmed theoretical; the measure almost certainly would have passed.

But before it could, the crown stepped in. As the primary customer of Virginia leaf, and a chronic sufferer of exchequer malaise no matter who the occupant of the throne, England was fond of low-priced imports, which were not only delighting smokers but doing wonders for the reputation of Charles's successor, James II. Some Brits thought the prices were the king's doing rather than the results of American surplus, proof of the monarch's craftiness in dealing with the unsophisticated colonists. James did nothing to discourage the impression.

In fact, he made every effort to maintain it, refusing to allow the Assembly to consider cutting back on leaf production, or even to meet. There would be no talk of tampering with supply, of artificially reducing yield. When the Assembly tried to convene anyhow, determined to pass its resolution, the crown's troops blocked the way. The Virginians were forced to disperse.

Word of this interference soon reached the plantations, and the owners, who had expected something of the sort, were ready with their response. They ordered their slaves into the fields—not to plant, but to sabotage; not to nurture, but to destroy. The slaves did not understand so perverse an instruction, but of course they obeyed. They marched through their masters' acreage and tore off leaves, uprooted plants, allowed weeds and worms to have their way. In some cases they burned the tobacco that was already being cured, and in others they threw stacks of cured leaves into the trash. The same thing happened in Prince Georges County, Maryland, where disgruntled growers were motivated by their "despair of any relief from the legislature," and where slaves, afraid that some kind of horrible trick was being played on them, their loyalty tested in a perverse and incomprehensible manner, kept looking over their shoulders, fearing retribution for the destruction they wreaked on the crops.

And so it went, not only in the seventeenth century, but through much of the eighteenth, crown and colonies continuing to disagree about the same issues that had caused disagreement from the start: the price of tobacco, costs of shipping, percentage of the costs to be borne by the Americans, inspection procedures, and other matters. To the kings of England, the very existence of tobacco in the New World sometimes seemed an act of defiance, a political statement of the most incendiary kind. They could not, or would not, see the leaf as a small, literal pleasure for a man or woman in search of a few moments of peace.

Its frustrations notwithstanding, though, the crown kept the trade going, kept the product coming. By the mid-1770s, the exchequer was taking in the hefty sum of 300,000 pounds sterling a year in export duties from the colonial planters, every single one of whom thought the amount excessive. Parliament, though, did not. Neither did the British citizens it represented, perhaps in part because they themselves were heavily taxed for a variety of goods.

But Edmund Burke, a newcomer to Parliament who would one day be his country's pre-eminent political thinker, sided with the Americans, persuading his fellow lawmakers not to provoke them further. Burke's "basic view," according to Theodore Draper, "rested on a distinction between the ideal and the practical." The ideal, for England, was more money. But the practical would likely be an American revolt, which in the long run would mean not only less money but less allegiance. As things turned out, Burke's prudence delayed the revolt by no more than a few months.

When it came, with the firing of shots at Lexington and Concord, Massachusetts, on April 19, 1775, Americans were startled as well as prepared, heartsick as well as eager. In a flash of gunpowder, the world became a different place, neither as familiar as it had once been nor as sensible. For one thing, colonial trade with England officially ended, and that with other European nations was severely reduced because of British embargoes. No more export duties for the colonists, but no more exports, either.

On the brighter side, or so some people seemed to think, was the increased availability of leaf for domestic consumption. Because of the war, the appetite was greater than ever, especially among men in combat. They smoked as a means of relieving—or, at least, distracting themselves from—the strains of battle. "Tobacco," Richard Klein has explained, "functions not to numb soldiers but to steel their nerves and to permit them to master the ambient anxiety that is their permanent condition."

It certainly had that effect on General Nicholas Herkimer. Stationed at Oriskany in upstate New York, where he commanded units totaling 800 militiamen, Herkimer was under orders to halt the British march across the state—not necessarily to kill, not even to wound, just to keep the enemy away by whatever means possible.

Everything was going fine for the general until the day he found himself ambushed by a combined force of British and native tribesmen. He was as surprised by their numbers as he was by their sudden appearance, but he did not flinch. "Although a ball had shattered his leg and killed his horse," writes Robert K. Heimann, "Herkimer continued to command his troops while smoking his pipe." And he commanded them skillfully. The British could not defeat the Americans at Oriskany and were soon in full retreat across all of northern New York. The general, one assumes, had another pipe, and puffed it with deep satisfaction, as he watched them depart.

It was to be expected, then, that when civilians asked George Washington how they could best contribute to the war effort, he said, "If you can't send money, send tobacco." Shipments were received eagerly by American soldiers, and when some expected leaf did not arrive on schedule, the men wrote plaintive letters home, as wounded in spirit as if sweethearts had spurned them. There are accounts from the time of soldiers smoking whenever a spare moment arose, enjoying the occasion not just for its own sake but for the memories it evoked of all the pipes and cigars they had savored in better days. And there are other accounts of men so greedy for tobacco, so disappointed that none had been sent to them, that they begged their mates to share, at times even came to blows.

Sometimes a fellow would pick up the cigar of a fallen comrade, lighting it as a kind of memorial, honoring his valor as well as satisfying the smoker's own cravings. Sometimes he would keep the pipe of the deceased as a souvenir. At night it must have seemed that the primary purpose of a fire was not to cook the meal or warm the body but to keep the tobacco aflame. It had been that way for a long time. "You stink of brandy and tobacco," says one character to another in William Congreve's 1693 play *The Old Bachlelor*, "most soldier-like."

Those on the home front also felt the war's emotional toll. They lit up day and night as they wondered about the outcomes of battle, the safety of loved ones, and the lives they would lead when the fighting finally stopped. Pipes and cigars became the companions that husbands and sons no longer were. Planters grew as much tobacco as they could, not worrying about prices. Once again the leaf was spent in addition to being smoked, not only paying the salaries of the fighting men but purchasing food, clothing, and occasionally arms. It might even have supported a spy or two at various times, although this is not certain. Regardless, a plentiful tobacco crop had become a patriotic imperative.

The enemy knew this. The enemy also knew that, as the war gradually wound down, quantities of American leaf began to find their way around British blockades to other European nations, providing the colonists with both financial resources and a renewed resolve to fight until freedom was assured. It was one of the signs that they were winning, proof to both themselves and the more perceptive among their antagonists that the end was drawing near.

The latter were furious. They tried to strengthen their blockades but did not have the resources—not the ships, the men, or even, at this point, the will. So in time they took a different approach. Under generals Phillips, Arnold, and Cornwallis, the British fought what historians have called, too grandly, the "Tobacco War," which resulted in the British seizing and destroying some 10,000 hogsheads of weed in Virginia in 1780–81.

Some of it belonged to Thomas Jefferson, who found its torching a "useless and barbarous injury." It was also the last gasp of a floundering army, an action that had no effect on the larger struggle at all. The "Tobacco War" was an expression more of pique than of sound military strategy. By this time, it was clear that the British had been defeated by their colonies and would soon have to recognize them as a nation of their own.

But the Americans were losers, too. During the years when none of their tobacco had been obtainable overseas, the English had turned to Turkey and Egypt, where they found not only acceptable substitutes for the colonial product but varieties of leaf that some began to prefer to the North American. They were at least as mild, at least as smooth on the intake, and for several years a lot more accessible. The Turkish and Egyptian blends became to many British the habit that *tabacum* had been before the war.

Other European countries, with American imports so greatly cut back, also turned to new markets. None of these countries, once the fighting ended, resumed trade with the United States at anywhere near previous levels. Only Germany and Russia continued to prefer the taste and price of American tobacco, but exports to those two lands did not make up for the business lost with others.

Thus, the final reason for the transformation of the Southern economy: With the war behind them, Americans were growing less tobacco because they were sending less across the ocean. But they were smoking more than ever themselves, thanks both to behavior that was reinforced in battle and to an increasing population. They were smoking as they conceived and implemented their new government and as they pushed westward to extend its reach. They were smoking as they built new towns and rebuilt old lives. They were smoking in homespun and in garments made of cotton from those wonderful new machines in the factories up North.

They were smoking despite the ever increasing doubts of one of their most eminent countrymen.

FIVE

Rush to Judgment

BENJAMIN RUSH, who appeared in the companion volume to this book, *The Spirits of America: A Social History of Alcohol*, now pays a visit to the present pages, and for much the same reason. The colonies' leading foe of demon rum, Rush became, as well, its first serious opponent of tobacco. As a physician, I wrote that Rush was "the leading figure of his era," known to many as the "'Hippocrates of Pennsylvania,' even though some of his ideas, like those of even his most distinguished contemporaries, were so misguided as to be counterproductive. He believed, for example, that certain diseases, such as yellow fever, were best treated by bleeding the patient." Bleeding them *so* much, in certain cases, that they eventually drifted into a blessed state of unconsciousness.

For this reason, modern historians—at least, those who bother to acknowledge him—often ridicule Rush, failing to see him in his time and thereby diminishing his accomplishments. They also fail to credit him with skills and ideas which seem as sound in the twenty-first century as they were radical in the eighteenth. For instance, Rush insisted that a physical ailment could have mental or emotional causes. He worked for the more humane treatment of mental illness when none of his peers thought such a course sensible. In less controversial pursuits, he had as long and prestigious a résumé as virtually any of the other remarkable Americans of the eighteenth century: surgeon general of the Continental Army, signer of the Declaration of Independence, cofounder of Dickinson College, member of the American Philosophical Society, abolitionist, public sanitation advocate, congressman.

No less impressive was Rush's list of friends. There was Benjamin Franklin, who did not smoke; Alexander Hamilton, who wanted to tax tobacco after the war to support the needs of the growing nation; and James Madison, who explained to Hamilton that such a tax would be antithetical to the principles of the revolution, as it would fall most severely upon "the poor, upon sailors, day-laborers, and other people of these classes, while the rich will often escape it."

Benjamin Rush's other friends included George Washington, although he felt that the great man was sometimes too full of himself, and Thomas Jefferson, with whom he exchanged fascinating letters on politics, religion, and the general state of society. And his friends were John Adams and Patrick Henry and James Otis, Paul Revere and James Monroe, Robert Morris and Philip Livingston. In fact, Rush's friends were almost all of the men who provided the ideological underpinnings for independence, as did he himself.

When few Americans had heard of Thomas Paine, Rush was calling as much attention to him as he could. In 1776, Paine published his fiery pamphlet *Common Sense*, which not only urged the break with England, but demanded a strong federal union afterward. Rush made sure his fellow patriots noticed, talking up the pamphlet and later writing that it burst on the scene "with an effect which has rarely been produced by types and paper."

The same could not be said, to Rush's chagrin, about an essay of his own. Called "Observations upon the influence of the Habitual use of Tobacco upon Health, Morals, and Property," it was the first anti-smoking tract of note in the United States. In it, Rush charged that the weed was particularly harmful to the mouth, stomach, and nerves, and like some Old Worlders before him, he thought he knew where tobacco would lead. "One of the usual effects of smoking and chewing," he wrote, "is thirst"—a thirst that "cannot be allayed by water"; that can, in fact, only be satisfied by "strong drinks." He went on:

> One of the greatest sots I ever knew, acquired a love for ardent spirits by swallowing cuds of Tobacco, which he did, to escape detection in the use of it; for he had contracted the habit of chewing, contrary to the advice and commands of his father. He died of a Dropsy under my care in the year 1780.

Rush was equally opposed to pipes and cigars, and his condemnations were the unofficial start of the American anti-smoking movement. Very unofficial: There were no organizations at first, no meeting houses, not even members in any kind of formal sense. There were just random individuals in random places who believed that the good doctor was on to something, or who had come up with similar ideas themselves independent of Benjamin Rush, perhaps without even having heard of him. How could a substance that looked so dirty be good for you? they asked one another. How could something that smelled so foul not be injurious? How could a plant that tasted so bitter not cause some form of distress? And, perhaps, a question from those of hellfire-and-brimstone religious inclination: How could something that brought so many people so much pleasure possibly be good for the soul? In the latter part of the eighteenth century, tobacco was blamed in some quarters for almost every ailment that had not obviously been assigned to something else: hemorrhoids and impotence, epilepsy and rheumatism, paralysis and headaches, perverted sexual practices and a high crime rate, and a "black, loathsome discharge from the nose," among other things. As the title of Rush's essay reveals, there were also concerns that tobacco contributed, through means not entirely understood, to a weakening of the moral fiber and the loss of materialism's fruits.

Dr. Joel Shew, a physician of the early nineteenth century and a disciple of Rush, started counting the various forms of illness and bodily breakdowns caused by tobacco, and did not stop until he had reached eighty-seven. Other doctors, though, were more restrained, even reluctant, about accusing the weed of wrongdoing. There was simply no evidence, they thought. Suspicion, yes, but nothing that would stand up in a court of law or, more to the point, in a laboratory. Medical science was not very good yet at matching causes to effects—not the correct ones, at any rate—and even if it had been, those who practiced it might not have wanted to speak against a custom as popular as smoking. They might have feared an icy reception. They might have feared a loss of patients.

In addition, as was true in other cultures as well as at other times, many doctors were putting out the opposite word, that tobacco was a kind of medicine, and it was a notion that would have a long life span. One day, George Eliot's Silas Marner would take up a pipe because his physician believed it was "good for the fits," and even if it were not, "that it was as well to try what could do no harm." A few decades later, Mark Twain's Huckleberry Finn would rely on the same device to settle his nerves. "I set down again, a-shaking all over," he confessed, "and got out my pipe for a smoke." And Twain himself, who claimed that he "came into the world asking for a light," further explained that he made it a habit "never to smoke when asleep, and never to refrain when awake."

Just as those who distrusted tobacco charged it with all manner of crimes against well-being, so did those who believed in it prescribe it for many of the same maladies.

FOR THE MOST part, Rush's support came from clergymen who did not live in tobacco country, men who stood behind him as preachers of the gospel had stood behind the leaf's foes in Europe. They manned their pulpits and orated passionately, making abstention from tobacco seem divinely ordained. But in only a few cases did they address its contributions to disease; instead, their primary objection to smoking was its relationship to alcohol, which Rush viewed as but one offense among many. Booze was the true culprit, most of the clerics believed, and tobacco but its nasty accomplice. Reverend Orin Fowler: "Rum-drinking will not cease, till tobacco-chewing, and tobacco-smoking, and snuff-taking, shall cease." Reverend George Trask: "Tobacco and alcohol were Satan's twin sons." Dr. R. T. Trall: Tobacco and alcohol were "the great-grandparent vices."

In the 1820s and 1830s, the temperance movement was getting up its first head of steam in the United States. It had started, however haltingly, half a century earlier with another publication of Rush's, this one called *An Inquiry into the Effect of Ardent Spirits*. In it, Rush was even harder on liquor than he had been on the weed. He believed that liquor would kill a person, and that such a death could only be considered suicide:

> Yes—thou poor degraded creature, who are daily lifting the poison bowl to thy lips—cease to avoid the unhallowed ground in which the self-murder is interred, and wonder no longer that the sun should shine, and the rain fall, and the grass look green upon his grave.

But few people heard Rush in the early days of nationhood, and fewer still put down their glasses or stayed home from the taverns.

Fifty years later, though, the time was right. Alcohol abuse was rampant in the land and acknowledged as a curse to the entire society. As a result, new temperance groups were being formed; older ones were adding members and shedding their labels as laughingstocks; and temperance literature was rolling off the presses faster than it could be read. A recurring theme was that Americans had freed themselves from the King of England; now they must free themselves from alcoholic beverages, no less vile and demanding an oppressor. They had freed themselves from the Anglican Church; now they must resist the Satanic influence of demon rum. There were public demonstrations to convert the reluctant, and those who agreed to take the pledge were sworn in on the spot, raising their hands and reciting oaths to refrain evermore from imbibing the liquid fires of perdition.

If only cotton growers would take the oath. If only other white men who so callously enslaved the blacks would repeat the vows. If only they would give up drink and clear their minds and unburden their hearts, they would surely see the error of the ways. This kind of thinking became an article of faith among those in the temperance battalions: Eliminate alcoholic beverages—or, at least, the abuse of them—and you go a long way toward eliminating human bondage.

It did not, of course, hold up to careful examination. But there were few careful examiners. Temperance and abolition became partners for a time and shared a zeal that was unique in both kind and quantity. They also shared members, shared speaking platforms at civic and religious events, and pooled their resources for a number of common goals. It was a union, thought a young lawyer from Springfield, Illinois, a few years later, that proved especially beneficial to the dry forces. Because of it, the temperance cause no longer seemed "a cold abstract theory," said Abraham Lincoln, but "a living, breathing, active, and powerful chieftain, going forth 'conquering and to conquer.'"

Alcohol's opponents clearly had momentum on their side. Tobacco's, however, were struggling, and they knew in their hearts the reason: The weed still appeared to be free of consequences. People did not throw up immediately after smoking too much; they did not lose their balance or get the shakes or become suddenly argumentative. They did not wake up the next morning with throbbing heads and steel-wool mouths. They might cough a few times while enjoying their pipes or cigars, but that did not seem indicative of deeper problems. It took a while for these to develop, sometimes many years, and by then there was no way to know how or why they had started. So the anti-smoking movement, such as it was, had no choice in the early to mid-nineteenth century but to rely on instinct, the gut-level revulsion of its members to the weed, and to ride for the nonce as far as it could on temperance's coattails.

It did not get far, even after a promising start. In fact, well before the Revolutionary War, laws were passed to restrict smoking, with Connecticut being the most zealous of the colonies. In 1646, public smokers in New Haven were fined six pence for each transgression. After a time the penalty shot up to eighty-four pence, "which is to goe to him that informs and prosecutes." A year later, the Connecticut General Court ruled that no one "under the age of 20 years, nor any other that hath not already accustomed himself to the use thereof" could partake of tobacco without a note from a doctor stating that it was "useful for him." Then the court got it into its head that a traveler could smoke only if he were on a journey of at least ten miles and only if he limited himself to one cigar or one pipe. The court did not say who would do the measuring or counting. By now, the colony's legal code was putting smokers into the same category as "common idlers and those hunting birds for mere pleasure."

More than a decade earlier, the court of Massachusetts had decreed that tobacco was to be a solitary pleasure; people were no longer to smoke in groups, even groups of two, either publicly or in private. That seemed to give the authorities in the village of Plymouth an idea. They decided to forbid the consumption of tobacco within a mile of a dwelling and to outlaw it completely in farm fields and rooms at inns.

In New Amsterdam, the dreaded Willem Kieft, known even to those who liked him as "William the Testy," went further. As director-general of the city, he banned the use of tobacco altogether, perhaps the first European to have done such a thing in North America. No one was to smoke either a pipe or a cigar, no one was to sniff or chew, at any time or for any reason within city limits. No exceptions. The illogic of inhaling hot and filthy vapors, or of snorting or chewing a vile weed, must at last give way to reason.

The response of New Amsterdam's residents to the edict is described with obvious relish by Washington Irving:

> The populace were in as violent a turmoil as the constitutional gravity of their deportment would permit—a mob of factious citizens had even the hardiness to assemble around the little governor's house, where settling themselves resolutely down, like a besieging army before a fortress, they one and all fell to smoking with a determined perseverance, that plainly evinced it was their intention to funk him into terms.

And funk him they did. It was the world's first smoke-in, and it worked. The ban was lifted; tobacco regained its previous standing in the city that would one day be New York; and Willem Kieft, without doubt, got even testier.

What happened in New Amsterdam happened elsewhere, just as certainly, if not as promptly or dramatically. By the end of Benjamin Rush's life, none of the previously existing laws against the weed was still in effect, not a single one. And anti-smokers were about to be dealt even more blows—blows from which it would take generations to recover—by the citizens of an expanding nation that seemed to require the expanded use of tobacco as a condition of its growth.

SIX

Ghost, Body, and Soul

YOU COULD TELL a lot about a man from the way he took his tobacco. You could tell what he thought about himself, what he wanted others to think about him, and how far he was willing to go to be so identified. You could tell about his personality, his standards, his style. And, increasingly, you could tell a lot about a woman.

The pipe had been the first smoke of the American colonists, in large part because it had been the first smoke of Englishmen. Reinforcing their preference were their observations of the native tribes of the New World. The Europeans found, or thought they found, that the more hostile types, those tribesmen less welcoming to the white man, smoked muskets. Crude smoke, crude people. The more sensible natives, meaning the kind of tribal leaders who would listen, negotiate, bend a little for their new neighbors from across the sea—these men chose a pipe, one that was almost certain to have been made of clay. "Wealthy aristocrats and royalty sometimes boasted of silver pipes," Jan Rogozinski tells us, "but these were mainly for show. Some gentlemen and burghers owned porcelain pipes ornamented with paintings or pipes carved out of fragrant woods. However, since porcelain is fragile and wood burns up quickly with steady use, these were too costly for most consumers. Meerschaum, a soft mineral that looks like white clay, also is fragile as well as expensive."

That a pipe was more trouble than a cigar might also have been a reason for its appeal. It made smoking a ritual rather than just an offhand activity, and rituals are a sign of an advanced—or, at least, advancing—civilization. And so the perception would remain, with the pipe taking precedence over the cigar, the clear choice of Americans of all ages until many decades later.

It was with a man named Israel Putnam, "Old Put" to his friends, of whom there were many, that things began to change. No one knows Old Put today, but he was a genuine American folk hero: five feet, seven inches of rock-solid, muscle-rippling, sweat-spewing manhood, the subject of one tall tale after another, a few of which might even have been true. Shortly after the Revolutionary War broke out, for example, Old Put is supposed to have driven a herd of sheep all the way from Connecticut to Boston, and all by himself: 130 animals, 100 miles. He had been told of food shortages among the American soldiers, the story goes; he wanted to make sure there was as much meat as possible on every fighting man's plate.

Within days of arriving in Boston, he decided to become a fighting man himself. He served the cause of independence with bravery and cunning, rising through the ranks so quickly that some of those with whom he served became jealous, talking of him behind his back, plotting his downfall. Old Put shrugged them off. Possessed of character, he was better known for being one. A Hessian who met him in Princeton, New Jersey, thought him brash, crude, and capable. He decided that Old Put "might be an honorable man, but only the Americans would have made him a general."

Some years earlier—in 1762, to be exact—Old Put had been fighting *for* the British, not against them. The French were the enemy, the Seven Years' War the venue. Shipped off to Cuba, and none too happy about it, Old Put found himself suffering from heat, fatigue, and, perhaps more than anything else, poor companionship. Like other Americans on the side of the Motherland at this time of worsening relations, he constantly had to endure

> the disdainful disregard of the British officers, and one day one of them had challenged him to a duel. It was Putnam's privilege to choose the weapons, and he chose to have each of them sit on a keg of gunpowder furnished with a slow-burning fuse. The first man to be blown up would lose. The kegs were brought and the matches lit. Putnam sat with folded arms, his nose in the air, his hat cocked over one eye, his heels casually kicking the barrel staves. The Englishman, on the other hand, "disturbed by the approach of the flame and his unusual situation," suddenly jumped off the barrel and withdrew to a safe distance in disgrace.

But no less frustrating to Old Put than the attitudes of British soldiers was the unavailability of his favorite pipe blend. Here he was in Cuba—to be specific, in the *Vuelta Abaja*, some of the richest tobacco-growing land in the world—and because he could find nothing that was shredded finely enough to pack into the bowl at the end of his stem, he could not enjoy a smoke. As a result, he could not unwind from the rigors of combat. Or keg sitting. He was cranky and getting crankier, and longed more and more each day for his beloved

Connecticut farm and the kind of weed he savored. Finally, one of his men insisted that he try a native smoke, a cigar, latter-day kin of the muskets that Columbus and his men had seen in that part of the world, but a bit more refined.

Old Put demurred, not wanting to be unfaithful to the pipe that had been his companion for so many years, not trusting something that seemed so radically different. But when a companion-in-arms reminded him that it was either a cigar or abstinence for the duration of the fighting—and who knew how long *that* would be?—Old Put changed his mind. He took one of the smokes, worked it reluctantly between his lips, and attached a flame to the end. He drew the smoke into his mouth, felt it infuse him, felt himself warm. So far, so good. After a few moments he let the smoke back out, slowly and attentively, watching the dark cloud drift away. In that instant, he became a convert.

The first thing that struck him was the taste. It seemed tangier than that of his pipe, more masculine, and thus it fit better with Old Put's image of himself, as well as with the military activity in which he was now engaged. The more Old Put thought about it, the more he realized that a pipe did not lend itself to war; it was too cultured, too suggestive of leisure and peaceable circumstance. He also liked the aroma of a cigar. Robust in itself, it did double duty by masking the noisome odors around him: death and disease, filth and despair.

And a cigar felt good to Old Put, surprisingly good. It was big and round and satisfying to the hand; it took up just the right amount of room and its shape allowed the fingers to configure themselves in a most comfortable fashion. There was always a certain awkwardness to a pipe, Old Put now realized, and this was especially inappropriate in battle, when a man needed to get his smoke going quickly and keep it going, often on the move.

Perhaps more than anything else, though, Old Put liked the sensation of having his lips right down there on the leaf. It seemed … truer somehow, more of what the experience of smoking should really be like. He did not drink his beer through a straw; how could he have spent so many years sucking tobacco smoke through a cold, hard pipe stem?

Everything about the cigar, in other words, set it apart from the pipe. And that, in turn, gave Old Put and others yet another way to set themselves apart from the British. It was an important concern to Americans at the time: Had they emulated the crown's troops, in however unintended a manner, they would have displayed subservience. A cigar was a small symbol of the gathering drive toward independence, just as was the preference for certain kinds of alcoholic beverage rather than the tea so often associated with the Motherland.

Men wanted to set themselves apart from women, too. Well before the colonies separated from England, Hannah Pemberton, "of the topmost rung of Philadelphia Quality," was smoking a pipe both at social gatherings and on the street. Her friends were incredulous at first, then followed her lead. In time, at least within her own circles, Pemberton's endorsement was all the cachet a pipe needed.

Other women in other places also took up the pipe "without the least attempt to conceal it, or the least apparent sense of its indelicacy." Some of the women were, like Pemberton, of "Quality"; others resided lower on the social scale. All, though, thought the pipe a clever gadget and relished the mild blends of tobacco and the fruity, sweet-scented flavorings that were available. Eventually, such respected individuals as Dolley Madison, Mrs. Andrew Jackson, and Mrs. Zachary Taylor would be known as pipe smokers. It was not a common practice among women at the time, nor would it ever be, but it was common enough to push at least a few males toward cigars, which the female of the species employed far less frequently.

WHEN THE SEVEN YEARS' War ended, Old Put got on with the rest of his life. He loaded three mules with all the Cuban cigars they could carry and marched them across the *Vuelta Abaja* to the docks at Havana. Unpacking the beasts himself, not trusting his cargo to the hands of others even for a few moments, he toted the smokes onto his ship and kept them next to his bed as he sailed north to home. When he left his quarters for a stroll on deck, it is said, he locked the door behind him.

Back in Connecticut, contentedly plowing his fields and grazing his livestock and delighting in the absence of Englishmen, Old Put blew out puffs of gritty, manly smoke that curled around his head and surrounded him like a squad of guardian angels, keeping him safe and blessed. He introduced his friends to cigars, his friends introduced them to their own friends, and Israel Putnam encouraged them all. He was the first American importer of fine Havana cigars and did for that smoke in his country, although on a smaller scale, what Sir Walter Raleigh had done for the pipe in England many years before.

After the Revolutionary War, Americans began to make their own cigars, with commercial production getting under way in South Windsor, Connecticut, in 1801. The domestic smoke was not nearly as good as the foreign, its taste coarse and in some cases overpowering, the proper methods of blending and curing for a cigar proving difficult for the newcomers to master. Even so, the U.S. version offered some of the same tactile pleasures and connotations as the Cuban and provided them at a fraction of the cost. Soon a small company in eastern Pennsylvania, near the town of Conestoga, was putting out something called a "stogie"—"a black, twisty, cheap and strong" cigar that became synonymous, at least among those who could not afford better, with good times and a vigorous outlook.

Literary monuments, which are the most enduring kind and which had been erected to the pipe in England, would not be erected to the cigar for many years yet, but when they finally came, they would appear all over the world, not just in the United States. Turgenev reveals the courage of a character in his short story "A Fire at Sea" by telling us that, under the title conditions, the man "quietly smoked his cigar and surveyed us each in turn with an expression of mocking pity." Stendhal wrote that, "On a cold morning in winter, a Toscan cigar fortifies the soul." Both Kipling and Lord Byron composed poems to the smoke; the former has his narrator lament the fact that his fiancée, Maggie, sees the cigar as competition for his affections.

Light me another Cuba;
I hold to my first sworn vows,
If Maggie will have no rival,
I'll have no Maggie for spouse.

And then there was poor, stolid Charles Bovary, so lacking in graces. His "conversation was as flat as a street pavement, and everyone's ideas but his own promenaded there in all their humdrum dress, bringing no emotion to his face, no smile, no look of contemplation." His wife came to disdain him for these reasons and many others, among them his ineptitude at handling a cigar, which he did clumsily. He "pouted out his lips, kept on spitting and drew his head back every time he puffed." Madame Bovary could not bear to watch; her husband with a smoke in his hands was so typical of her husband without one.

On the battlefield, though, it did not matter how a man handled his tobacco, only that the weed be there for him. In his classic Civil War novel *The Red Badge of Courage*, Stephen Crane writes of the moment before a regiment, closing in on its enemy, "went swinging off into the distance":

> As the horseman wheeled his animal and galloped away he turned to shout over his shoulder, "Don't forget that box of cigars!" The colonel mumbled in reply. The youth wondered what a box of cigars had to do with war.

A box of cigars also figures in Willa Cather's *A Lost Lady.* The title character's husband, Captain Forester, asks a woman in his company for permission to open his box and take out a smoke. She gives it. Then he asks another woman. She, too, assents. "Had there been half a dozen women present,"

Cather writes, "he would have asked that same question of each, probably, and in the same words." Shortly afterward, a game of whist begins with an air of gravity and a subtle revelation of character, neither of which would have been possible, Cather implies, without the civilizing effects of tobacco.

Becky Sharp, in William Makepeace Thackeray's *Vanity Fair*, was also willing, even eager, to have a man light up in her company:

> "You don't mind my cigar, do you, Miss Sharp?" [asked Captain Rawdon Crowley.] Miss Sharp loved the smell of a cigar out of doors beyond everything in the world—and she had just tasted one, too, in the prettiest way possible, and gave a little puff, and a little scream, and a little giggle, and restored the delicacy to the Captain; who twirled his moustache and straightaway puffed it into a blaze that glowed quite red in the dark plantation, and swore—"Jove—aw—Gad—aw—it's the finest segaw I ever smoked in the world, aw," for his intellect and conversation were alike brilliant and becoming to a heavy dragoon.

As a student at Bowdoin College, Nathaniel Hawthorne, who would one day create fiction both eerie and unforgettable, "resented regulations stipulating how far one could walk on the Sabbath and that forbade smoking a 'seegar' on the street or consuming alcohol."

Even the ashes of a cigar came to be prized by some. When "mingled with camphorated chalk," it has been written, they "make an excellent toothpowder; or, ground with poppy-oil will afford for the use of the painted a varied series of delicate grays."

It is true that others, in real life as well as in fiction, did not care for the cigar. It is true that long before Freud, jokes were being made about its phallic significance, and some people wondered whether this was the real reason for Becky Sharp's elation. And it is true that the esteemed newspaperman Horace Greeley would one day dismiss the cigar as "a fire at one end and a fool at the other." None of it mattered. The cigar began to gain on the pipe in popularity the moment that Old Put sailed home from Cuba, and it would continue to gain until, three decades into the nineteenth century, it surpassed the pipe among Americans, the latter being too pretentious, too inconvenient, and too British ever to catch up again.

Even a few women, after a time, turned to cigars. They seem especially to have caught on among the women in the mining camps during the California Gold Rush. And, as J. D. Borthwick relates, they simultaneously found favor among members of the fairer sex in Panama, some of whom ended up in those mining camps as well as in other mid-century American locales:

> They have a fashion of making their hair useful as well as ornamental, and it is not unusual to see the ends of three or four half-smoked cigars sticking out from the folds of their hair at the back of the head; for though they smoke a great deal, they never seem to finish a cigar at one smoking. It is amusing to watch the old women going to church. They come up smoking vigorously, with a cigar in full blast, but, when they get near the door, they reverse it, putting the lighted end into their mouth, and in this way they take half-a-dozen pulls at it, which seems to have the effect of putting it out. They then stow away the stump in some of the recesses of their "back hair," to be smoked out on a future occasion.

But to the chagrin of smokers, non-smokers, and spittoon cleaners alike, another form of tobacco would in time be more coveted than either the pipe or cigar and would prove—in symbolic terms, at least—an even greater break with England.

BACK IN THE GOLDEN age of exploration, there were sights that Europeans would rather not have seen, places they would rather not have gone—for instance, Isla de Margarita, a large landmass off the coast of Venezuela at the turn of the sixteenth century. Most adventurers from the Old World avoided it. But Amerigo Vespucci stumbled into the vicinity one day and made his way ashore. He discovered some natives indulging in a "green herb ... to such an extent that they scarcely talk." However, they were not smoking it; they were chewing, treating the plant as if it were food, a vegetable, of all things—and most historians believe that this is the oldest of all forms of tobacco use, preceding smoking by several centuries. Even the Mayas were known to have chewed on occasion, mixing their tobacco with lime.

Vespucci and his men did not rank among the more genteel of God's creatures, yet even they were appalled by the sight of human beings chomping on tobacco leaves, working up a big, foaming mouthful of expectorant and then hawking it out and letting the excess run down their chins. They thought they were looking at animals, creatures of an incomprehensible sort.

Vespucci would have been even more appalled if he had seen the Isla de Margarita tribe, and others like it, in combat. In addition to their usual weaponry, they were

> primed for conflict with a generous plug of tobacco. Warriors, at close range, would then seek to blind an enemy by squirting tobacco juice into

his eyes. This tactic demanded furious mastication by the spittle squads, the explosive force of a guided missile and superb marksmanship.

A similar technique was aimed at Spaniards when they landed in Paraguay in 1503; the natives were "chewing herbs and spurting the juice towards them." Actually, they were scoring direct hits, splattering the Spaniards' faces and stinging their eyes and fouling their clothes. The invaders dropped back in confusion and disgust, wiping themselves as best they could and cursing. The natives watched them retreat, sighing in relief and hanging on to their lands for a few more years. Apparently, these tribes began chewing tobacco as a means of allaying hunger on long journeys. Only later did they see its potential as an armament.

Most Europeans, once they got used to the idea of tobacco in the first place, preferred to smoke it. But a few took up chewing, especially sailors, who replaced their pipes with plugs of leaf when they crossed the ocean, thereby eliminating the risk of fire on a wooden ship. Plugs also kept their hands free so they could enjoy their weed at the same time that they did the ship's work.

There were chewers, as well, in colonial America, but not many, and most were of the lower classes. John Hancock was an unfortunate exception. Despite the best efforts of Benjamin Rush and other physicians, Hancock was thought to have died from ailments brought on by his fondness for chewing tobacco, swallowing the putrid juice because he was too much of a gentleman to spit it away. American chewers also needed their hands free, and they generally worked in the kinds of places, such as farm fields and blacksmith shops and, later, factories, where they could discharge their tobacco fluids without staining property or offending finer sensibilities.

But it was with the presidency of Andrew Jackson, a rough and ready man, that chewing tobacco became a national pastime, the man and the era collaborating perfectly, if inadvertently, to further the cause of hard-pumping jaws and dark brown spittle. Jackson occupied the White House from 1829 to 1837, and it may be said of the era that neither the eagle *nor* the tobacco plant was the nation's ideal emblem. It was the spittoon. True, Jackson, like his wife, also smoked a pipe, "a great Powhatan bowl with a long stem," but as his nation was now a plug kind of nation, so was he a plug kind of leader.

Jackson could not be called an educated man, not in the conventional sense. "He had vigorous thoughts," it has been written, "but not the faculty of arranging them in a regular composition." Nor did he have, in the conventional sense, a president's upbringing. He was the first American head of state who had not emerged from either the Virginia or Massachusetts elites, who was not, in fact, even couth as these people understood the term. They would not have invited him to dinner. They did not vote to send him to the White House.

Born in the Carolinas, the child of poor farmers, Jackson actually lived for a time in a log cabin. His father died a few days before he came into the world, his mother and favorite older brother when he was fifteen. By this time, Jackson was a drinker and a gambler and something close to a hooligan, a young man as quick-tempered as the part of the new nation that had sired him:

> Brawling in barrooms, sporting with young ladies, moving outhouses in the hours well past midnight—such activities gave Jackson a reputation as "the most roaring, game-cocking, horse-racing, card-playing, mischievous fellow that ever lived in Salisbury [North Carolina]," according to one resident.

It was the temperament of a man-at-arms, and after seeing limited action against the British in 1776, Jackson went on to become a hero in the War of 1812 and a villain—at least in the eyes of future generations—as an Indian fighter. With the latter role, Jackson became obsessed, and in the most cold-blooded fashion. He seemed personally affronted that so many natives occupied so much good farmland and even after being elected president—perhaps especially after being elected president—wanted nothing more than to get rid of them and turn the land over to white folks.

In fact, during his years in the White House, and at his insistence, Congress passed at least ninety treaties calling for the removal of indigenous American tribes from their lands, and he saw to it that more than half a million dollars and a large if unspecified amount of manpower were appropriated to carry them out. When, on one occasion, the Supreme Court overruled Jackson and refused to allow the Cherokees to be herded away from their ancestral homes in Florida and Georgia, he dismissed the decision as "too preposterous" and, in what Alistair Cooke calls "one of the most shameless and arbitrary acts of an American President," Jackson defied the court and ordered the army to evict the natives forcibly.

"And so," Cooke goes on, "in what is truly called 'the trail of tears,' thirty thousand Cherokees were persuaded or chained, gently led or viciously driven, as far west as Oklahoma, and along the way a quarter of them died." The causes of death were "dysentery and pellagra, cold and hunger, brutality and despair. Both the rich and the poor suffered, especially women and children." It was a ghastly episode. The tears should have fallen down the soldiers' faces no less than their victims'.

But there was another side to Andrew Jackson: more compassionate, more moderate, as kindly disposed toward his own kinds of people as he was belligerent and hard-hearted toward others. He fought for labor unions and government regulation of banks; he stepped up trade with England and handed out federal jobs to friends as if he were Santa Claus and his bag were bulging at the seams. He was a sandy-haired boy with freckles grown to crude, rumpled adulthood, and if he liked you, he wanted very much for you to like him back.

He was "the original populist," as George F. Will has called him. No wonder his supporters were thrilled, even vindicated, when he defeated the dignified John Quincy Adams for president, denying him a second term. And if Jackson's people got a little too exuberant on Inauguration Day—if, in fact, the proceeding turned into "one of the most notorious scenes in American history"—well, at least the Jacksonians were more honestly expressive than those cold-blooded New Englanders and those nose-in-the-air Virginians.

> Chief Justice John Marshall administered the oath. Afterward, President Jackson mounted his horse and rode leisurely at the head of an enormous crowd of followers to the White House for a planned reception. The crowd thronged into the East Room. All order quickly dissolved. The people surged toward the bowls of rum punch. Windows shattered. Fights broke out. Noses were bloodied. Ladies fainted. The crowd of well-wishers pushed President Jackson out a rear door and forced him to take refuge in the nearby Gadsby's Hotel. Only a quick removal of the punch bowls and food tables to the lawn saved the mansion from further damage. Conservatives shuddered over what appeared to be the start of another French Revolution. "The reign of King Mob seemed triumphant," a fearful Justice Joseph Story recalled later.

The people—or the people to whom historians like to refer as "the people"—loved Andrew Jackson. They admired his mettle, supported his programs, and re-elected him in 1832 by 150,000 votes out of 1.2 million cast. When, after eight years, he was finished with the presidency, they voted into office his chosen successor. And they prayed for Jackson in retirement, for his struggles with tuberculosis and the loneliness of a widower's state that had begun just before he assumed the nation's highest position. In his final years at his home in Nashville, called the Hermitage, he would often sit of an afternoon in his favorite rocking chair, leaning slowly forward and back, thinking his private thoughts, chomping on the plug of tobacco in his mouth. Those who visit the Hermitage today will still see the spittoon on the floor beside the chair, conveniently situated if long out of use.

Andrew Jackson did not make his age; he merely stood for it, although so perfectly that the one cannot be imagined without the other. Yet the age would have been raucous even if old John Quincy had beaten him in 1828, or if the decent and eloquent Senate leader Henry Clay had prevailed in '32. For having fought the Revolutionary War to win its liberty and the War of 1812 to solidify itself, the United States was now turning its formidable energies to internal matters and in the process entering a rambunctious and belated adolescence. "If movement and the rapid succession of ideas and sensations constitute life," wrote a French visitor named Chevalier halfway through Jackson's first term, "in America you live a hundred-fold. Everything is circulation, mobility, and a terrifying agitation."

The Americans circulated west even more than they had before, far too restless to remain within old borders or even relatively new ones. They would have told you they were purposeful in their migrations: looking for fertile farmland, lush pastures, rich soil, roaring rivers, tall trees, taller mountains, rare metals, new outlooks, different opportunities, freedom from the past and all other forms of restraint—and so they were. But they were also circulating for the same reason that an animal moves when it has been too long confined—for the pure joy, if not the biological necessity, of motion. They were flexing muscles, spreading wings, going out on limbs.

They did none of it slowly. They ran instead of walked, walked instead of standing still. They whipped horses and raced wagons and sped down rivers in steamboats, sometimes with results that were literally explosive. They did not think it important to be first at something in particular; they wanted to be first at *anything*, for the sheer exhilaration of haste. And why not? Without haste, how would Americans cover all the land that was theirs, do all the living that the new country demanded by being so large and raw and welcoming?

They did not worry about manners. Etiquette was not sissified to Americans so much as time-consuming and irrelevant; they had no desire either to learn it or to practice it. And so they bumped into one another without excusing themselves, splashed mud on one another with barely an acknowledgment. They belched and passed gas in public; sometimes they urinated or defecated wherever they happened to be when the need came upon them and were not embarrassed if someone saw them. After all, that someone might have been doing the same thing himself not long before and been observed by other public relievers of themselves. A few people ate with their hands, almost no one with forks. To Americans, forks were "a foppish indulgence of the decadent Continental upper class."

Like those who had chewed tobacco before them, these Americans needed unoccupied hands. They had the machines of the Industrial Revolution to operate, as well as plows to steer, trees to clear, homes to build, oars to stroke, reins to pull, game to shoot, crops to raise, livestock to feed, tools to sharpen, and a host of other activities of one sort or another to pursue. Once implanted, a plug required no maintenance and only the effort it took to move the jaws every few seconds. And they were moving anyhow, constantly moving, as people bragged to their neighbors about how well they plowed and cleared and steered and built and all the rest of it, daring anyone to contradict their accomplishments. "It may be," says Jerome E. Brooks, who writes engagingly of the chewing-tobacco heyday, "that the exciting events of those times demanded that the jaw be kept well lubricated for the requirements of argument and oratory."

The plug was, in short, the perfect tobacco for people who were constructing a nation from scratch. With spittoons or open spaces enough, the budding country's work would get done both promptly and efficiently, if not altogether hygienically.

In symbolic terms, the plug was also apt, perhaps even more so than the cigar. Americans had finally made their peace with England, but only in the sense that battles were no longer being fought. If the British were not now political oppressors, they remained cultural oppressors, or so it was believed by many in the United States who resented British art and fashion and style and, for that reason, wanted to keep their distance more than ever, just as adolescents want to pull away, once and for all, from their parents.

Noah Webster helped with the distance. He was publishing his spelling books, grammars, and readers and phasing out the Motherland's English in the process: no more *u* in humor and labor, for example; no more *re* in center and theater.

Tavern owners helped with the distance. They were changing the names of their establishments from those of British origin to those honoring the victors' Revolutionary War heroes: from the King's Arms or the Crowing Cock to the Lighthorse Harry Lee.

Authors helped with the distance. The United States was developing a literary class—or, at least, a few literary individuals—and they were creating distinctly American characters having distinctly American experiences in distinctly American settings. The characters created by James Fenimore Cooper and Washington Irving were talking Andy Jackson's language, not the Buckingham Palace variety.

Reverend Samuel Francis Smith helped with the distance. In 1831, he wrote a new set of lyrics to "God Save the King." His version was known as "America" and became an unofficial anthem.

"Europe," writes Neal Gabler of this period, further explaining the importance of distance to those on the other side of the Atlantic, "was everything

America was not and should not be: effete where America was earthy, refined where America was natural, intellectual where America was practical, decadent where America was moral."

And pipe smoking where America was plug chewing. From President Jackson's time to the Civil War, Americans seem to have chewed far more than they lit up, and over the course of the entire nineteenth century, chewing tobacco accounted for half of all the leaf manufactured in the United States. The 1860 census for Virginia and North Carolina, as an example, lists 348 tobacco factories, only seven of which were processing their product for smoking. Pipe blends had fallen off so much in the marketplace that they were sometimes put together from the leftovers of plug making, more a means of utilizing waste than of catering to consumer demand.

Some companies sold their chewing tobacco straight. Others flavored it with rum, nutmeg, tonka beans, sugar, cinnamon, licorice, and honey. The names were even more singular than the tastes. There were brands of plug called Live and Let Live, Buzz Saw, Barbed Wire, Bull of the Woods, Cannon Ball, and Beat the Dickens. With regard to the last, there is a story that begs to be told.

BY 1842, HE COULD WAIT NO LONGER. It was time to come to the United States, he told his friends, time to stop reading about it and imagining and see it for himself. Although barely out of his twenties, he was already the most famous living author in the English-speaking world, having produced *The Pickwick Papers*, *Oliver Twist*, *Nicholas Nickleby*, *The Old Curiosity Shop*, and *Barnaby Rudge* in the past six years. His books were particular favorites in America, in part because of the heart-on-the-sleeve appeal of their sentimentality, in part because of the disregard they showed for Britain's upper classes, the reigning taskmasters of government and capitalism, who, in turn, seemed to have so much disregard for the upstart Americans.

He was not searching for acclaim. It was material that Charles Dickens wanted from the land across the sea, his intention to write a nonfiction book on the former colonies that would satisfy not only his fellow Brits' inquisitiveness about them but his own. His was especially acute, for "he had long been optimistic about the American experiment. He firmly believed that in America under a republican system of government the daily practices of life would reflect the highest principles, whether in business, politics, or religion. There he hoped to find a freedom from the corruptions of ... social injustice, and discredited snobbery that characterized English society. America would be a signpost pointing the way for England to follow." He hoped, in other words, to find a place where the Oliver Twists of the real world were treated kindly by their elders, and the Nell Trents could live long, fruitful lives.

But it was acclaim that he got, in unprecedented amounts, at unprecedented volume, on display for him before he even walked down the gangplank of his ship:

> As the *Britannia* pulled into Boston Harbor, a dozen newspaper editors, "came leaping on board at the peril of their lives" to interview the international celebrity. Every town in America wanted the honor of entertaining him publicly, as if to authenticate its cultural status. Every newspaper wanted interviews, quotations, anecdotes, exclusives, publicity.

All of which made Dickens even more optimistic about what he would find in the United States, more confident that this was the place where his kindred spirits resided, not in the land he called home. "I can give you no conception of my welcome here," Dickens wrote back to England. "There never was a king or emperor upon the earth so cheered and followed by crowds, and entertained in public at spending balls and dinners, and waited on by public bodies and deputations of all kinds." In fact, we are told by James C. Simmons, "Not even the Beatles's reception on their first American tour matched the depth and intensity of feeling that Dickens generated across the entire spectrum of American society over a century earlier."

Dickens's first doubts did not arise until he had departed from Boston and headed south. One day, shortly after arriving in New York, he was riding with some friends in a wagon, the group of them conversing lightly, enjoying the sights and the bustle. Then, as they crossed Broadway, Dickens looked up and to the side. Conversation stopped. The author gulped. His friends followed the author's startled gaze. To the amazement of all, they saw that "a select party of half a dozen gentlemen hogs have just now turned the corner."

Maybe it was true, then, Dickens might have thought. Maybe America really *was* this untamed backwater that so many of his countrymen thought it to be. Maybe it was as coarse in its own ways as England was in others. Dickens did not want to believe it. But the evidence was about to accumulate.

A few weeks later, this time traveling in a canal boat through the western states, he found the sight of his fellow passengers even more dumbfounding than that of the urban swine. They had "yellow streams from half-chewed tobacco trickling down their chins." Yet it was not just on aesthetic grounds that Dickens objected to the weed munching; he also complained about the consequences for others, saying that at times it was almost impossible to sleep in an adjoining bunk or cabin or hotel room. "You can never conceive what the hawking and spitting is, the whole night through."

Perhaps Dickens tried to convince himself it was just this particular boat, just this particular part of the country. He soon learned otherwise. He saw chewing and spitting everywhere he went, indoors and out, at work and at play, among men and women and children, some of whom, he concluded, "expectorate in dreams." He saw more spittoons than any other piece of furniture. And he marveled at the pride that Americans seemed to take in their ability to hit the spittoon from almost any distance or angle. "Set right there, stranger," they were known to say to one another, pointing to a chair next to the receptacle, "an' I'll jest miss ya!"

Dickens never took such a chair himself. He had no confidence in the American aim, noting that "even steady old chewers of great experience are not always good marksmen, which has rather inclined me to doubt that general proficiency with the rifle of which we have heard so much in England."

On another occasion, Dickens had reason to rue the absence of a spittoon. He had invited a visitor to his hotel room one afternoon, and as the two of them began to talk, the man pulled out a wad of tobacco, inserted it between cheek and gums, and began working away at it. Dickens was not surprised, did not object. He had been in America long enough now to know the customs. But when the man looked around the room and saw no container for his waste, he decided that all he could do was hawk his juice out the window. He took a breath, reared back, let loose. Only problem: the window was closed. The spit hit the glass. The man either did not notice or did not care. He kept hawking and talking, talking and hawking, and ignoring the proliferating number of beige spittle streams as they trickled down the panes of glass and pooled on the sill.

The most depressing sight of all to Dickens was probably Washington, D.C., the seat of government in this nation that, at least from afar, he had admired so greatly. To his readers in England he wrote that

> Washington may be called the headquarters of tobacco-tinctured saliva. ... In all the public places of America, this filthy custom is recognized. In the courts of law, the judge has his spittoon, the crier his, the witness his, and the prisoner his; while the jurymen and spectators are provided for. ... In the hospitals, the students of medicine are requested by notices upon the wall, to eject their tobacco juice into the boxes provided for that purpose. ... In public buildings, visitors are implored, through the same agency, to squirt the essence of their quids, or "plugs" as I have heard them called by gentlemen learned in this kind of sweetmeat, into the national spittoons, and not about the bases of the marble columns.

Dickens expressed his disfavor time and again. The sights of chewing were as offensive to him as the sounds, the taste as offensive as the resulting behaviors. Taken together, they made up "the most sickening, beastly, and abominable custom that ever civilization saw." It is no surprise that a plug company wanted to "beat the author" in its brand name.

SEVERAL YEARS EARLIER, another British tourist in the United States had had similar experiences. Mrs. Frances Trollope, or Fanny, as people called her, was the mother of Anthony, who would one day be almost as famous a man of letters as Dickens. And Fanny herself rivaled Dickens in popularity for a time; she, like him, turned out books that were fueled by social conscience as well as literary passion.

Trollope came to America to be part of Nashoba, a utopian community in Tennessee that had dedicated itself both to the abolition of slavery and the resettlement of its former victims. But the location depressed her. "Desolation was the only feeling," she said about the backwoods South, "the only word that presented itself." After a time, the community failed, and the causes behind it seemed to be facing a similar fate. Trollope thought about going back to England but decided instead on Cincinnati. It was here, as she struggled for a time to make ends meet, that she gathered the material for her first and best-known book. It was titled *The Domestic Manners of the Americans.* She did not think highly of them.

In fact, Trollope might have been even more disgusted by the constant spitting of her subjects than was Dickens. She saw them spit in the streets and in restaurants, in theaters and in shops, on trains and in carriages. In the drawing rooms of Cincinnati's finest homes, she noted, "the gentlemen spit, talk of elections and the price of produce, and spit again." Boarding a steamboat on the Ohio near the end of 1828, Trollope looked down at the carpet and tried to muffle a shriek:

> I may not describe its condition; indeed it requires the pen of a Swift to do it justice. Let no one who wishes to receive agreeable impressions of American manners, commence their travels in a Mississippi steamboat; for myself, it is with all sincerity that I declare, that I would infinitely prefer sharing the apartment of a party of well-conditioned pigs to the being confined in its cabin.

Perhaps if she had crossed the same Broadway intersection that Dickens would later cross ...

After a three-year stay, Fanny Trollope finished her survey of the various behaviors of the American citizenry and wrote a letter to the editor of the *Cincinnati Enquirer*, previewing the tone that her book would later take. "This has been called the 'age of improvements,'" she said, "the 'age of novels,' the 'age of lectures,' &c. For my part, I think it is more especially the *age of tobacco chewers* more than of anything else."

It probably was, and would continue to be after Trollope went home and after Dickens went home and even after the departure of the Frenchman Alexis de Tocqueville, the most perceptive of all nineteenth-century European visitors to North America. Tocqueville admired much about the United States. What he did not like, among other things, was the number of its citizens who would "smoke, chew, spit in your beard."

All in all, writes J. C. Furnas, "No single other thing, not even Negro slavery, did so much as the stains on the American eagle's white shirtfront to encourage the supercilious European to label the new nation barbaric."

Which was, of course, a large part of the plug's appeal.

FOR THE OPPOSITE REASON, snuff never had a chance. It was too precious, too dandified, too ... *un*barbaric. Few Americans tried it and fewer still became regular users. A cigar you could taste and smell, a plug you could taste and feel, but snuff was tobacco in finely powdered form that you held in front of one of your nostrils and snorted up to your eyeballs, then held up to the other nostril and inhaled with similar vigor—and what kind of sense did *that* make? Your eyes would widen, perhaps lose focus, begin to water. You would shake your head and breathe through your mouth. After a few seconds you let out a sneeze that would have registered on the Richter scale, if there had been such a thing back then. You wiped your nose, smiled absently, and, somehow invigorated, started the whole process again. For what possible reason?

(There are actually two kinds of snuff, nasal and oral. The latter, as the name suggests, is packed between the gums and cheek and resembles plug, although the tobacco is usually of a finer cut and, at least in modern times, possessed of a gentler taste.)

It was Columbus, on his second voyage to the New World, who seems to have discovered snuff, and it confounded him as much as the smoke eating he had seen the first time around. One of his shipmates, the Spanish monk Ramon Pane, described the natives' method of taking it in. "Snuffing is through a tube," he later wrote, "one end placed over the powdered leaf and the other in the nose, and so drawn up, which purges them very much."

It is thought that Columbus brought some of the powdered leaf back to Spain with him, another curiosity. This one, though, did not catch on. The pleasures of smoking made themselves apparent to great numbers of people after a relatively short time; snuff remained, to most, a mystery. For more than a hundred and fifty years, it was little used in the Old World, and almost no one wrote in its behalf.

Then, starting late in the seventeenth century, the mystery was solved, at least to the satisfaction of a few people. If smoke is the ghost of tobacco and plug the body, as has been said in modern times, snuff was now recognized by its newfound aficionados as the soul. As such, it became the sudden rage among European nobility, the overnight vogue at one court after another. The gentlemen and ladies of the royal precincts would finger it, inhale it, talk about it as ceremoniously as if it were an affair of either state or the heart. So frequently was snuff sniffed in these circles that the first edition of the *Encyclopedia Britannica*, published in 1771, defines it as "a powder, chiefly made of tobacco, the use of which is too well known to need any description here."

The attractions, as the court denizens reflected on them, were several and probably should have been recognized earlier. Snuff was clean; it required no unsightly spitting, produced no cloud of smoke, no offensive odor. In fact, snuff actually smelled good, at least when it was blended, as was the case in most European countries, with perfumes or flowers, aromatic herbs, or wines like port and Madeira.

Thus, it had another attraction: A blueblood with a snout full of snuff was a blueblood protected from the rank scents around him, so many unwashed people practicing filthy habits in unsanitary surroundings. Prior to snuff, he had had to stuff a perfume-soaked piece of silk into his nostrils and breathe deeply. It was a lot of trouble and not a very comfortable way to stroll around town.

Sometimes the powder even looked as good as it smelled, having been dyed with a red or yellow pigment. Like plugs, snuff was safer than pipes and cigars; there was no chance of a fire that would scorch jewels or consume paintings, destroy tapestries or ruin royal robes, or singe long and swooping mustaches.

Before too much time passed, and perhaps inevitably, people were claiming that snuff had medicinal properties. Said a writer of the time, tobacco for the nose

> heals colds, inflammation of the eyes, involuntary tears, headaches, migraines, dropsy, paralysis, and generally all those misfortunes caused by the pungency of the humours, their too great amount and their dissipation from their normal conduits. Nothing is better to increase the fluidity of blood, to regulate its flow and circulation. It is an unfailing sternutatory to

> revive those with apoplexy or those in a death trance. It is a powerful relief for women having the pains of childbirth; a certain remedy for hysterical passions, dizziness, restlessness, black melancholy, mental derangement.

And as if all that were not enough, snuff was the perfect accompaniment to royal snobbery, for the motions that propelled the powder into the sinuses—the curling of the tip of the nose, the flaring of the nostrils, the pushing back of the upper lips and cheeks—were the very motions upon which aristocrats had long been relying to convey the scorn for the masses. Taking snuff was, in other words, a means of keeping the facial mechanisms of contempt in good working order.

People became attached to their powder, strongly attached. They sniffed it all the time, at virtually any hour of the day or night. Sometimes, unable to bear to part with the stuff, they left it in their noses while they slept. Not a good idea. This practice was "found to occasion vomiting usually on the next morning. Another thing charged on this way of application, is, that it weakens the sight."

But to the swells at European courts there was more to snuff than merely ground leaf. There were also the vessels in which it was kept. These were prissy little works of pretentious art with a practical purpose, and they seem to have contributed as much to snuff's appeal as the granulated tobacco itself.

At first, people stored their snuff in leather pouches or small, unadorned wooden boxes. But these catchalls were too unassuming for a habit of such ostentation. The search began for more suitable materials, and soon snuff boxes were being fashioned from the most treasured substances that the world had to offer:

> The mineral, vegetable and animal realms have responded bounteously, both in their simplicity and in combination. All the minerals are represented, from platinum, gold and crystal to lead and iron. The vegetable products are represented by almost every kind of wood, bamboo, gourd, and amber. The lower animal life yield all sorts of shells, the queen of which is the pearl oyster, and its rival for the throne, the amphibian tortoise. The higher animals have furnished leather, bones, horns, tusks, and the king of all animal products, ivory, which is not only beautiful in itself but takes colours with brilliancy, and has challenged by its fine grain the skill of artists and engravers since the early Egyptian dynasties.

But once the earth provided the raw materials, it was up to craftsmen of the highest skill to fashion them into the proper shapes, and then to artists of the highest skill to provide the decoration, to paint and engrave and in other ways lovingly adorn. Among the scenes rendered on snuff boxes of the eighteenth century, in fine line and poignant colors, were woodlands, seascapes, the heavens above, palatial residences below, and men and women without a stitch of attire copulating like deranged bunnies set loose in elegant boudoirs. The artist Richard Cosway, whom history remembers chiefly because Thomas Jefferson wanted to cuckold him, was known, among other things, for producing snuff boxes of extremely erotic design.

It was up to the partakers of snuff to see to it that they had a number of containers in their possession. "One has boxes for each season," writes Louis Sebastian Mercier of eighteenth-century Paris. "That for winter is heavy; that for summer light. It's by this characteristic feature that one recognizes a man of taste. One is excused for not having a library or a cabinet of natural history when one has 300 snuff boxes."

If, that is, one knew how to operate his box. Withdrawing it from pocket or purse and then opening and closing it and putting it away—these seemingly mundane steps became in time a social grace apart from the use of its contents. They called attention not just to the powder inside but to the fingers that handled it, and these were invariably laden with glittering rings: diamonds and emeralds and rubies and all manner of other precious stones, flawlessly set and often so large that the digits supporting them bent inward, toward the palm, from all the weight.

But how was a person to learn this grace? One way was for an aspiring dandy to emulate a practiced one, to serve a kind of apprenticeship through observation and interrogation. Another was for him to enroll in an institution, a charm school of a sort, that devoted itself to snuffing's protocol. There actually were such places in Europe a few centuries ago, and they never wanted for students, or for excessively qualified faculty.

According to what seems to have been the most commonly accepted curriculum of the time, the snuff box was to be extracted from purse or pocket with the left hand. The lid was to be tapped a few times with the fingers of the right—an attention-getting device, one assumes, although also a means of loosening the powder. Then the box was to be opened and a tiny amount of snuff placed on the back of the left hand, or perhaps between the thumb and index finger. Following this, one brought the powder to the tip of the nostrils, paused a moment so that one's rings could blind an onlooker, and huffed the snuff as far back into the nasal cavities as it would go.

But not everyone was so earthy, so low-down. Some people came positively undone at the notion of touching powdered tobacco with their epidermal layer, possibly even getting some of it under their nails. To avoid such a calamity, they carried little spoons around with them, also works of art, these from master silversmiths and exquisitely shaped and engraved and worth far too much money. Often the nobles wore them on chains around their necks, constantly at the ready, like the cocaine addicts among the beatnik crowd at a later time in America.

Such fellows might also equip himself with a little brush, or even a hare's foot, to whisk the excess powder away from his upper lip. One account, which may or may not be true, has it that the noted fashion plate and infamous ladies' man Beau Brummel invented the handkerchief for the express purpose of dusting specks of snuff from his waistcoat.

It should surprise no one that Old Put preferred a fat Havana, that Andy Jackson kept his spittoon right by the side of his rocker.

There were few criticisms of ground tobacco at the time from the upper classes. One of them came from the Englishman Lord Stanhope, who objected to it not because of its pointlessness or conceit, not because of its excess or bizarre ritualism, but because of the time wasted in consumption. "If we supposed this practice to be persisted in for forty years," the gentleman calculated, "two entire years of the snuff-taker's life will be dedicated to tickling the nose, and two more to blowing it."

Lord Stanhope did not factor in the amount of time spent sneezing, but he should have. Sneezing was, without question, the most controversial aspect of snuff. The ladies and gentlemen at court might have enjoyed a good, deep whoosh—in fact, the literature of the period suggests that they found the experience almost orgiastic—but the lower classes were of a wholly different mind. They feared, as a result of sneezing, "that their sense of smell would fade, their heads would fill with soot, and their brains would dry out."

Others believed that the force of a sneeze might loosen vital organs, disturb internal processes, perhaps even blow a person's soul right out of his body. This is why they said "God bless you" immediately after an eruption and why they looked at snuff not as a substance for personal gratification but as something that the prince of darkness had whipped up in the lab to further his nefarious ends.

The pro-sneeze faction found this absurd. In their view, snuff rid the body not of its spiritual essence but of its "superfluous humours." And who needs superfluous humours? They also insisted that snuff brightened the eyes, whetted the appetite, and generally improved one's outlook on life, even in the midst of trying times.

To Jerome E. Brooks, as astute about the powder as he was about the plug, snuff and the various theories on sneezing made the nose the subject of more discussion and more tributes in the eighteenth century than ever before in recorded history. Before the custom of snuffing had run its fantastic course, a person was being envied for a generous nose and expansive nostrils as much as for a handsome face or a good figure.

Not in the United States, he wasn't. Not in the land of the free and the home of the cigar and plug. Here noses remained what they had always been, passages for oxygen and a few other assorted gases and scents, not for fine, flavored grains of tobacco. Early in the nineteenth century there was a factory or two in the Northeast that made snuff, but the product never attracted many customers, and most of them treated it as a luxury, something for a special occasion as opposed to an item of daily routine. A few others saw something medicinal in snuff, including Ben Franklin, who took a pinch now and then as "a cure for the Hickups." For thirty years, Franklin claimed, he and his friends made the annoying ailment go away within seconds of its onset with a simple application of the nose powder and some hearty inhalations.

There were also clergymen who would load up their nostrils before a sermon because of snuff's inconspicuousness, thus enabling them to take to the pulpit and rail at the malefactions of tobacco without being outed as hypocrites. Eventually snuff would appeal to those in monasteries, mines, and other places where it was unsafe to smoke and unseemly to spit. It appealed to those among the democratic hordes, always a minority, who fancied themselves aristocrats at heart, and, in the mid-nineteenth century, the oral variety of snuff seems to have found favor for a time among women:

> The female snuff-dipper takes a short stick, and wetting it with her saliva, dips it into her snuff-box, and then rubs the gathered dust all about her mouth, and into the interstices of her teeth, where she allows it to remain until its strength has been fully absorbed. Others hold the stick thus loaded with snuff in the cheek, a la quid of tobacco, and suck it with a decided relish, while engaged in their ordinary avocations; while others simply fill the mouth with the snuff, and imitate, to all intents and purposes, the chewing propensities of the men.

Other than that, Americans took their tobacco in forms that better suited their nature, especially between 1861 and 1865, when the very meaning of being an American was called so violently and tragically into question.

IN THE STORY of *Tobacco in America*, Joseph C. Robert writes that there are three reasons men smoke more than usual in wartime. They are "(1) absence of family restraints, (2) indulgence by way of escape from the fatigues of military life, and (3) quickened imitativeness accompanying the massing together of people." In other words, they smoke because (1) no one tells them not to, (2) they are under a lot of pressure, and (3) everyone else is doing it. All of these factors were present, and to great extent, in the Civil War, which was the bloodiest and most emotionally wrenching conflict in American history and, not incidentally, the end of the old plantation system for both tobacco and cotton.

But smoking was promoted by the locations of battle as much as by its exigencies, as armies fought on fields where tobacco was grown and found lodging on farms where it was cured. On occasion, Southerner introduced Northerner to the weed, both cigar and plug; a few times, the combatants even smoked or chewed together during breaks in the conflict—small, privately arranged truces, details of sanity in the big picture of aggression.

Not only was there, as Robert points out, an absence of familial restraint during the war, there was the presence of military encouragement. Commanding officers wanted their men to smoke, knowing that they needed distraction from the ennui and horrors around them and much preferring tobacco to booze. A man who smoked too much could still aim his gun and hit the enemy; one who drank to excess might pull the trigger and amputate his toe.

The war's most famous smoker was Ulysses S. Grant, the rousingly successfully Union general and future president who was also the war's most famous tippler. When a temperance group, revolted by tales of Grant's imbibing, stormed the White House to demand his dismissal, President Lincoln met with them but was not sympathetic. "Doctor," he is reported to have said to the group's leader, "can you tell me where General Grant gets his liquor? ... for if you could tell me, I would direct the Chief Quartermaster of the Army to lay in a large stock of the same kind of liquor, and would also direct him to furnish a supply to some of my other generals who have never yet won a victory."

His drinking might have been exaggerated; his smoking was not. Grant seldom mounted a horse, addressed his men, or even stepped out of his tent for some fresh air without a cigar tucked between his lips, wiggling up and down as he chewed on the end. "No caricaturist who drew Grant without a cigar in his mouth," writes John Bain Jr., "could hope to rise in his profession."

After a while, Grant did not even have to buy his own smokes. He was given more cigars than he could ever use himself, despite his prodigious appetites, by people who were grateful for his military prowess. Often he would go into battle with his smoke unlit, waiting until he was certain of victory before striking a match, and then he would pull on the tobacco so hard, so many times, so quickly—he would, in other words, surround himself with such a swirling profusion of blue-gray smoke—that he looked like a person who had been set afire, a victim of the stake rather than a triumphant warrior.

In fact, so much did Grant and the other fighting men of the North value their leaf that, near the end of the war, Southern troops, who were no less fond of tobacco's consolations than their foes, decided to punish them for it. They set fire to Richmond, their own capital, almost destroying it because it was also the center of American tobacco production; to the victor would not go these particular spoils. It was the Confederacy's version of the "Tobacco War" and was as meaningless and desperate an act as the British equivalent had been many decades before. It has been reported that Richmond was still aflame when General Robert E. Lee surrendered to Grant at Appomattox to end the fighting.

Tobacco also served as currency in the Civil War, although not to the extent that it had in earlier times. Some men were paid with it, some supplies purchased; a few sheaves were held in reserve in the event of financial emergency. Occasionally the weed was promised as a bonus, a way to persuade a soldier to give extra effort or an official at a storage facility to release additional materiel.

But it was not by substituting for cash that tobacco made a significant financial contribution to the Union side; it was by bringing *in* cash, as the federal government decided for the first time to put a direct tax on the products that men smoked. Factories were assessed five cents per pound of manufactured goods in the last months of the war, a figure that shot up to forty cents as the fighting came to a close. Consumers were also taxed: four cents for each batch of five-penny cheroots and four cents on each pricey nickel cigar. Chewing tobacco seems to have escaped a levy, except for a small one placed for a time on the manufacturers. The first full year of the tax was fiscal 1864, and the government, to its own surprise, took in 1.3 million dollars.

By 1957, to pick a year at random, the total was more than two billion dollars.

But there was another tobacco tax during the Civil War, a tax that could not have been charged in previous conflicts because the item in question was not available in those days, at least not in this country. Smokers had to pay a surcharge of forty cents for each thousand of these items at first, and then in the war's final year, two dollars per thousand—and they railed at the increase. They cursed the government, cursed the tobacco farmers, cursed the companies that processed the weed. What they did not do was abstain. In fact, they would have paid even more to avoid abstaining. In the years to come, they *did* pay more, much more, and they did so, however reluctantly, because they did not see themselves as having a choice.

The item was called a cigarette, and it would change the habits of Americans—and, in fact, people all over the world—like no other tobacco product in history.

SEVEN

The Cigarette

WHEN THE NATIVE tribes of the Americas made their tobacco muskets, they did not use the entire plant. They cut off the stems and damaged leaves and sometimes allowed little corners of leaf to fall to the ground in the process of shredding or packing. They were later discarded, left to rot or be eaten by animals. As the natives did not consume every piece of the plants they ate, neither did they smoke every piece of the plants they inhaled.

But in time, the tribesmen began to see the practice as wasteful, possibly worse. Tobacco was a gift from the gods, after all; to throw away even a little of it was to reveal an ingratitude that might be kin to heresy. They asked the gods' forgiveness, believed that it was granted, and promptly mended their ways.

But they would not go so far as to add the leftovers to their muskets or sprinkle them into the bowls of their pipes. Instead, they collected them and mashed them into a powder, although not so fine a one as snuff. Because they were working with small quantities here, they divided the powder into smaller portions than those that made up the muskets. Then they wrapped each of them in a piece of reed or bark or straw or sometimes a leaf of maize or a leaf from a tree or even a banana skin, packing the tobacco tightly and smoothing off the ends. They had no name for these little smokes, and not much regard for them. They gave them to the less discriminating members of their tribes: the very young, the very old, the poor, and women.

These tobacco dregs, bundled in tubes and handed out like alms, perhaps never even witnessed by Columbus and his immediate followers to the American continent, were the world's first cigarettes.

CENTURIES LATER, in preindustrial Europe, cigarettes were reinvented, again as the smoke of outcasts. Scrounging through piles of garbage in the cities, paupers found cigar butts, snuff remnants, and crusts of tobacco that the well-to-do had scraped out of their pipes and thrown away. They collected these rejects in a container of some sort, ground them into a rough kind of tobacco gruel, and wadded it into pieces of paper they had also found in the garbage. They put these homemade smokes in their pockets or special hiding places and went about their business, whatever it was, until sunset.

Then they came together, gathering in alleys and doorways and other shadowy places with the rest of their kind, out of sight of decent folk, and lit up. They puffed slowly, having nothing else to do until dawn. They felt the ghost of tobacco inside them, prowling the corridors of their bodies, altering their perceptions and distracting them from their woes. It was as close as they ever came to pleasure, yet it was only that. The homeless of the Middle Ages never dared to dream that the smoke they released on earth would catch the eye of a sympathetic god.

THE FIRST FIRMS to manufacture cigarettes did so in Seville, Spain, sometime in the seventeenth century. The little cylinders of tobacco were still nothing special, "a poor man's by-product of the lordly cigar—scraps of discarded cigar butt wrapped in a scrap of paper." It was cigars in which the companies specialized; they made cigarettes simply to cut down on rubbish, the same reason that some American plug companies would later manufacture pipe blends.

But as time passed, more and more people began switching to cigarettes. The new smokes were easier to light than pipes or cigars, and easier to keep lit. Their improvised blends of tobaccos and smaller size made them surprisingly mild, a special attraction for women and men of more refined tastes. And a cigarette could be consumed in a fraction of the time demanded by a pipe or cigar; a woman could smoke one between household tasks, a man between rows of the field he was planting. The cigarette, in other words, could accommodate a person's schedule, rather than the other way around. The compactness of the new smoke made it an innovation.

So it was that, starting in Spain, the humble little offspring of the musket made its way across Europe, eventually becoming established on both sides of the Atlantic, ascending the social ladder sometimes before 1800 and moving to Portugal, Italy and South Russia. In Brazil, it was called papelito; in Spain papalete or cigarillo; in Italy, where the product was larger than we know it, a paper cigar. The French monopoly, or Regie, began its manufacture in 1843 and the word "cigarette" is of French origin.

It was not until the middle of the nineteenth century that the English began smoking cigarettes in large numbers, and it might have been even later had it not been for the Crimean War, where British soldiers picked up the habit from the Turks and Russians. The war was not a shining moment for the British: "The general professional incompetence displayed by officers of all ranks must surely have been fatal against any other European power than Russia." So abysmal, in fact, was the performance of British officers in the Crimea that training procedures for them would be thoroughly revamped shortly after the war ended, the war having established in irrefutable detail how *not* to engage a foe in combat. In the short run, the English might have thought cigarettes the only good to have come out of the whole experience.

However, it was a Scotsman named Robert Gloag who thought so first. Returning from the Crimea, he set up a tobacco factory in England, where he turned out "cylinders of straw-coloured paper into which a cane tip was inserted and the tobacco filled in through a funnel." They caught on immediately, the first smoke in that part of the world to be favored almost equally by men and women.

But it was in Spain where the little muskets, having had a head start, truly thrived. Several companies were producing them by this time, not only in Seville but throughout the land, and holding them in ever higher regard. They had begun to combine good grades of leaf with the scraps, and then better grades, and finally they eliminated the scraps altogether. People of both genders smoked them here, too, as well as people of all ages, all occupations, and all previous preferences in tobacco. Spain might have been a monarchy, but cigarettes were proving an amusement with republican appeal and soon became a more common sight in Barcelona and Madrid and Cadiz than pipes or cigars had ever been. According to a Spanish proverb of the time, "A paper cigarette, a glass of fresh water, and the kiss of a pretty girl will sustain a man for a day without eating."

THEY WOULD NOT do so in America—not at the start, at least. The girl would have to do more than kiss the guy; the glass would have to hold something a lot stronger than water; and the cigarette would provide no sustenance, or even enjoyment, at all.

It might be a quick smoke, but that only meant it delivered less enjoyment than its competitors. It might be easier to handle than the others, but that only meant it lacked heft and, to some, seemed hardly worth the bother of holding, of elevating to the lips. And a cigarette might taste mild, but in a nation that boasted of its hardy, devil-may-care vigor, this was not exactly a virtue.

To some Americans, in fact, it was a vice or, at least, a sign of weakness. When the men who dug the Erie Canal saw one of their fellows smoking a cigarette, their first impulse was to question his virility. Their second was to throw him into the canal. But things would change. Momentous events were ahead for the United States and in their wake would come new and surprisingly flexible attitudes toward tobacco.

THE HISTORIAN Robert Sobel believes that "war and advertising have always been the two most important methods of spreading cigarette use." As yet there was no advertising to speak of, but there was the Civil War. Cigars were the most common smoke for men on both sides of the conflict, not only because of Grant's example but because a soldier usually prefers the comfort of old habits, and most fighting men went back a long way with their cigars, as did a certain number of others with their pipes.

A few, though, were willing to experiment, and to them cigarettes seemed so tailor-made for battle that they might have been designed by a quartermaster. They were easier to carry than other tobacco products, taking up less room in a man's pack than cigars and weighing less than a pipe and a pouch of blend. A fellow might long for a leisurely smoke, but there was seldom time; an exchange of gunfire, a few minutes of puffing, another exchange of gunfire—this was the pattern of the soldier's days, and while most men just stuck with their cigars, sometimes relighting the same one several times, keeping it in their mouths both lit and unlit for hours at a stretch, other men were stoking themselves with cigarettes.

The latter, apparently, were the better tonic. It has been reckoned that a fine Cuban cigar of the period had a nicotine content of about 2 percent, and pipe mixtures and plugs even less. By contrast, the Virginia tobacco that went into cigarettes, whether ready-made or hand-rolled, might have as much as 9 or 10 percent. Thus, even though cigarettes had a milder taste, they had more of a kick, which was exactly what a soldier needed to fortify himself as his fellow warriors dropped from bullets or disease and he wondered whether he would be next. "When one smokes," Richard Klein speculates,

> one does not merely suck a tit of consolation; cigarette smoke is not always, not often, perhaps never mother's milk—it mostly tastes bad, produces a faint nausea, induces the feeling of dying a little every time one takes a puff. But it is the poison in cigarettes that recommends them to the heroic. ... In every puff there is a little taste of death, which makes cigarettes the authentic discipline of good soldiers.

But perhaps they were *too* suited to war. Potent instruments on the battlefield, agents of respite and rejuvenation, cigarettes went back to being effeminate little doodads when the fighting ended. Rather than bringing the custom home with them, many soldiers seemed to believe that they had now earned the right to return to their cigars or pipes, to smoke unhurriedly again in safe, familiar surroundings. To these men, cigarettes might have brought back too many memories, all of them the wrong kind. The *New York Times* even fulminated against the smaller smokes, saying in an editorial as late as 1883, "The decadence of Spain began when Spaniards adopted cigarettes, and if this pernicious practice obtains among adult Americans the ruin of the republic is close at hand."

The adult Americans among whom cigarettes first obtained in the post–Civil War years were probably women. In fact, there are a few reports, though only a few, of women sampling the leaf in this manner before the war. "In 1854," writes George Tennant, "an observer of New York smoking practices reported that 'some of the *ladies* of this refined and fashion-forming metropolis are aping the silly ways of some pseudo-accomplished foreigners in smoking Tobacco through a weaker and more *feminine* article which has been most delicately denominated *cigarette*.'"

Not many products of the time were made with women specifically in mind. Cigarettes seem to have been one of them, with their taste and shape and speed of consumption. Perhaps some members of the fairer sex were even titillated by a perceived naughtiness. According to Sobel, just the act of purchasing cigarettes "produced a sense of guilt and excitement [in women], and the actual smoking, the thrill of tasting forbidden fruit."

Richard Klein goes further. "A woman smoking in public," he writes, "offends those who think that women are supposed to be veiled. In private, between a man and a woman, it becomes the permanent signal of her determination to resist his male opinions, his will. ... Every puff she takes says to him that she is determined to take a breath, a puff, that is entirely her own."

Men were not fond of symbols of rebellion against them. Nor did they care for the fact that most cigarettes were imported and therefore had names that no self-respecting farmer or rancher or smithy or adventurer could say aloud without cringing. Think about it: Was he going to stride into the general store and slap his fist on the counter and demand some Opera Puffs? Was he going to stride into the local saloon and demand a whiskey neat and invite the bartender to join him in a Bon Ton or Cloth of Gold? Was he going to stock up on Entre Nous or Huffman Imperiales for the cattle drive? Even the word "cigarette" was a source of embarrassment, feminine and diminutive, not to be taken seriously by the kind of person who took his tobacco seriously.

Cigars, on the other hand, were often named after political heroes at home or testosterone-laden rulers abroad: General Grant, Daniel Webster, Henry Clay, King Edward, Emperor William. No shame attached to any of these. A man could clomp into a place of business and call the names right out and still hold his head up high.

All of which is to say that cigarettes had an image problem in the immediate aftermath of the Civil War, as other tobacco products had had, at various times among various peoples, before them. But an even bigger problem for the small smoke was means of production. Like cigars, cigarettes were made by hand, but it took five or ten of them to equal the smoking time of a single cheroot or stogie. Thus, even with a limited demand, a lot of cigarettes were required to keep customers puffing away for their accustomed intervals. How could a company possibly roll out enough merchandise to meet that demand? Where would it find enough employees? Or were human beings not really the answer?

THE LATE NINETEENTH CENTURY was a boom time for inventions in the United States, the adolescent nation having now reached young adulthood and begun to channel its energies more constructively than before, to seek more practical outlets. From the country's formal beginnings up to the end of the Civil War, a total of 62,000 patents had been issued to American citizens. Between then and the turn of the twentieth century, more than 500,000 were handed out, and they resulted in improvements in virtually every field of endeavor that mattered to a society: transportation, communication, education, nutrition, construction and engineering, mining, oil drilling, manufacturing, farm and household labors, clothing, medicine, hygiene, and entertainment. In the single decade following Lee's surrender at Appomattox, Americans gave birth to air brakes, the pneumatic rock drill, reinforced concrete, pressure cooking, bicycle production, the "duplex" telegraph, the electric voting machine, the stock ticker, celluloid, color photography, and the "zoopraxiscope," a crude version of the movie projector, to mention but a few advances.

And then, in 1881, to the surprise of tobacco men everywhere, there came into being a piece of equipment that turned cigarettes from an afterthought into an industry.

James Bonsack was an energetic sort, tireless, even fidgety. At the age of twenty-one, and something of a tinkerer by profession, he liked nothing more than to tinker through his leisure hours, as well. He was never happier than when putting machinery together or taking it apart, whether on a workbench, on a piece of paper, or in his dreams. He did not set out to revolutionize the tobacco business. There is no reason to think that he cared much about the leaf one way or another. He merely wanted to work with his tools, to create—and after much trial and an identical amount of error, he put together a Rube Goldberg-like contraption consisting of belts and pulleys, nuts and bolts, gears and gizmos. To look at a sketch of it today is to think of an erector set assembled by a child who has too many pieces at his disposal and no clear idea what to do with them.

But the apparatus was more efficient than it appeared, even a little simpler, as it "poured a flow of shredded tobacco through a feeder device and onto a thin strip of paper. The paper was rolled into a single long tube. As the tube came out of the machine, a rotary cutting knife cut it into equal lengths."

It was not the first such machine the world had known. But it was the best. "Bonsack's distinctive contributions were improvements in the tobacco feeding mechanism, the cigarette forming tube, and the cutting knife," writes Richard B. Tennant. "His machine weighed a ton, absorbed one-half horsepower, and required an operator and two feeders."

And because it could turn out a cigarette thirteen times faster than a man rolling his own, tobacco companies, most of which had already realized that cigarettes were the future of their industry, bought machines as quickly as Bonsack and his co-workers could manufacture them. James Duke, who founded the American Tobacco Company and later endowed Duke University with some of his profits, ordered a pair of Bonsacks. When they did not work as smoothly as he wanted them to, he brought in a mechanic to do a little fine-tuning. Then Duke put the machines back into service and made history. "On April 30, 1884," Arlene Hirschfelder relates, "the modern cigarette was born. That day the Bonsack machine ran for a full ten hours and turned out 120,000."

It was the output of fifty skilled hand rollers, maybe more, yet only three men, the operator and two feeders, were required to keep the machine running. The amount of labor needed to produce $1,000 worth of plug could now send to market $20,000 worth of cigarettes, which, unlike the handmade variety, were uniform in taste and appearance—quality control, even though the quality was not as good as it would eventually be.

Duke was elated, even though, in a sense, he was almost as indifferent to cigarettes as Bonsack had been. Duke never smoked them, never allowed the women in his family to smoke them. It was plug that he favored as a boy, cigars as an adult, pipes as a businessman in the company of his colleagues. But the pipe had not been doing very well of late, and Duke could not figure out what to do about it. He needed a different idea, some new product lines. So he took a chance on young Bonsack, and now that it was paying off, he wanted the world to know. On the packs of those new smokes he attached a label announcing their provenance. "These cigarettes," it read, "are manufactured on the Bonsack Cigarette Machine."

Other companies kept the automation a secret at first. They were afraid people would rebel if they knew that something as important as a tobacco product were being produced by something as impersonal, and as prone to malfunction, as a machine. But word got out, and no one seemed to mind. Machines were doing a lot of other jobs in nineteenth-century America that had previously been done by humans, including, in recent years, the manufacture of cigars. There was no reason for them not to turn out cigarettes, as well.

And turn them out they did, as the Bonsacks enabled America's tobacco companies to raise their output from 9 million cigarettes in 1885 to 60 million two years later.

The male smoker was even more receptive to machine-made smokes when the companies changed the names. Now he could ask for a Rough and Ready, accept a Wage Scale from a friend—he was comfortable with language like that. And when consumers looked at the price of cigars and pipe blends and plugs, they found the cigarette even more enticing. Thanks to Bonsack's invention, the small smoke could now be sold more cheaply than before, and more cheaply by far than any other tobacco product. The price went even lower as the federal government, no longer paying off the Civil War, reduced the tax on cigarettes, almost eliminating it for a while. This would have been an important consideration at any time; in these particular years, it was for many people the deciding factor in what to smoke or even whether to smoke at all.

BY THE EARLY 1870S, Congress had been providing the American railroads with free federal land for rights of way for two decades. The first transcontinental line had been completed in 1869, and hundreds of smaller lines were being laid all across the country, in all directions and at all angles. They were changing the landscape, changing the way people did business, changing the lives of the men and women who rode them to new homes, new jobs, new opportunities. They were an expression of raw power, these sets of metal tracks spreading across the United States. Sometimes they were used for the benefit of all, but too often, it seemed, they were used for the private gain of those who controlled them.

Hoping for even more of such gain, the owners of some railroads overextended themselves, building too much, too quickly, too irresponsibly, and borrowing too much money to finance it all. They became land-based pirates, looting their fellow owners and charging outrageous fees to shippers, trying everything they could think of to cut losses. They speculated in real estate, issued fraudulent stock, cooked up tall tales for creditors. Nothing seemed to work. Eventually, twenty-five U.S. carriers were forced to default on the interest payments for their bonds, and the country was plunged into a depression, the worst that Americans had known to that time.

It was called the Panic of 1873, and in one form or another it lasted almost until the Panic of 1884, the effects of which went right up to the doorstep of the Panic of 1893, when

> scores of railroads, hundreds of banks, and almost 16,000 businesses failed, and more than two million workers lost their jobs. The Sunday, July 14, 1895 edition of *The New York Times* contained only four notices in the "Help Wanted—Male" column, compared to sixteen times that number in the "Situations Wanted—Male" column. For women, no "Help Wanted" notices appeared at all, whereas more than ninety Notices for "Situations Wanted—Female" were listed.

All of this was caused, to put it simply, by men who rolled high, behaved low, and succeeded to an appalling degree in turning public marketplaces into private cash cows.

Some people survived the hard times, carrying on as they had before. Others had to adjust, and it was not easy. They sold their possessions, gave up their homes. Or, if they were not hit that hard, they began to make more of their own clothes, to eat cheaper foods and perhaps less of them than before, to cut back on amusements, to search for extra income wherever they could—and to smoke cheaper smokes. Even an inexpensive cigar cost two cents; a pack of ten cigarettes, though, went for a nickel, and if you rolled ten of your own, the cost was two or three cents.

Sales of the new product soared. Bad news for the nation was good news for the men with the Bonsacks. Rough and Readys and Wage Scales and the like still didn't seem quite right to a lot of American males, but they made do: the harsher the reality, the less important the image.

Cigarette sales also got a boost from matches, which, prior to the end of the nineteenth century, were expensive, brittle, and often dangerous, especially those whose heads contained white phosphorus or sulfur or both. "In fact, the ingestion of match heads in quantity," one observer noted, "was one way to commit suicide in this period." These matches were hard to light, requiring that a flint be struck against a piece of steel so that the resulting friction would create a flame, and so hazardous to produce that it was probably better for one's health to smoke cigarettes than to work in a match factory. What A. N. Wilson writes about London's firm of Bryant and May was true of many American companies, as well: "Phosphorus fumes filled the premises, and many employees—they were nearly all female—developed 'phossy jaw,' a form of bone cancer, or skin cancer. The hours were long—in summer, 6:30 A.M. until 6 P.M., in winter starting at 8 A.M."

But in 1892, a Pennsylvania lawyer named Joshua Pusey began making matches with heads of safer materials and sticks of cardboard instead of wood. He gathered them into something resembling the modern matchbook and hoped he would make his fortune. There was just one problem, and if there had been a *Saturday Night Live* at the time it would have made a wonderful skit. Pusey put the striking surface on the *inside* of the book. As a result, when a person meant to light but one match, he often burned up the entire book; the user's fingertips were as likely to catch fire as his cigarette.

It was, of course, an easy problem to solve. Soon large firms were manufacturing matchbooks with exterior striking surfaces, making a single flame possible rather than a conflagration. Matchbooks did not really catch on with the public, though, until a brewing company executive got an idea. Posterity does not seem acquainted with him, nor is it certain about the name of his beer, but the idea was to make matchbooks into advertising vehicles: load them up with slogans and promises and pictures of the products, then provide them to consumers free of charge.

Before long other brewers were doing the same. Then non-brewers. In a few years, free matchbooks were everywhere, promoting all kinds of goods and services, from exercise devices to patent medicines to household furnishings—and in the process encouraging Americans to light up more than ever before.

Urbanization, too, helped to sell cigarettes. In 1860, there were only nine cities in the United States with a population of 100,000 or more. By 1910, there would be fifty. In 1860, 20 percent of Americans lived in cities. By 1910, the percentage would be 45 and climbing steadily. At the same time, smaller towns were getting larger, and some of the smallest were either being absorbed by their neighbors or falling off the map. "Farmers who left the countryside for jobs in the city," we are told, "often turned to cigarettes on discovering that spitting was frowned upon in mixed company."

Once the farmers got to the city, they found that the leisurely strolls they had once so enjoyed, while not necessarily frowned upon, had become passé. Besides, there was no place to take them, as there were no places for an unhurried conversation, a sleepy afternoon, a quiet interlude. The pace of the large urban area was frenetic, like a runaway pulse beat. There were so many more people around these days, and so many demands on them. Factory whistles bleated; shop foremen snapped out commands. Men and women alike were suddenly faced with the demoralizing new concept of industrial efficiency, and it was the same thing as living under a ruthless monarch. City dwellers ran for trolleys, ate lunch standing up if they could eat it at all, lived and worked in the midst of buildings so large, so dominating, that the shadows they cast made escape seem impossible. It was as noisy at midnight as it used to be at noon. There was no time for a cigar, no time for a pipe, time only for a cigarette—one now, one a little later, one after that, whenever it could be squeezed into the day.

In 1869, 1.7 million of the small smokes were fired up in the United States. Six years later, the total was 42 million. New brands were appearing in stores every week, and reporters had begun to write stories that treated cigarettes not so much as a product but as the social phenomenon they were rapidly becoming.

Cigar makers began to fret. They feared the competition, feared that they would have reason to fear it even more in the future, and paid it an ultimate kind of compliment by reacting deviously, organizing a campaign of what we would today call disinformation. Back then, it was just lying. Through both word of mouth and printed page, they spread the message: Each cigarette that a person inhaled

> was drugged with opium and morphine; its paper container was bleached with arsenic and white lead; its contents were derived from stumps and "sojers" picked out of gutters by tramps and ragpickers; Chinese lepers were the chief producers of cigarette papers, and only suicidal people or idiots would smoke the noisome thing.

Hearing this, some smokers stayed with, or went back to, their cigars. Others, either ignoring the propaganda or simply not knowing of it, turned to cigarettes more than ever, hoping to calm their nerves in the jangled times in which they lived. Both forms of tobacco grew in popularity with a growing population. The cigarette was the workingman's salvation, the cigar a badge of the fellow whose time was his own to govern, or who wanted to give that impression. Cigarette: the shoemaker around the corner. Cigar: the guy who owned the block. Cigar sales, in fact, doubled in the United States between 1890 and 1920, then began a decline that would prove long and lingering, although not terminal.

But the plug's decline, which began gradually about 1880, *was* terminal, or the closest thing to it. Per capita consumption dropped from 2.8 pounds in 1890 to half a pound in 1937, and it was not just because of the city's better manners. Chewing tobacco, as Robert K. Heimann points out, "was a function of space and solitude. Space, because of the chaw's juice-generating character; solitude, because chomping was a substitute for chatter." Neither existed in the American city of the late nineteenth century.

It was not until much later, though, that tobacco-industry executives officially recognized the plug as a bygone commodity. On September 14, 1955, James Duke's American Tobacco Company circulated a memo to all directors and department heads. It read as follows:

> It has become impossible to hire persons in the New York area to clean and maintain cuspidors [spittoons]. Since the cuspidors presently on hand in the New York office can no longer be serviced, it will be necessary to remove them promptly from the premises. Removal will take place this weekend. Your cooperation in doing without this former convenience is appreciated.

Had she still been alive, had she not died a death so painful and so painfully ironic many years earlier, the indefatigable Lucy Page Gaston would have been pleased. More than that, she would have been vindicated. But she would have wanted more. Now, she would have thought, if only the American Tobacco Company and every other company in every other city engaged in the same damnable line of work would only remove *it*self!

EIGHT

The Carry Nation of Tobacco

SHE HAS BEEN described as "tall, ungainly, and rather bony, both in face and figure. She had a high forehead, large upper lip, and mouse-colored hair, which gave her a rather masculine appearance." In fact, she had the appearance of one man in particular: Abraham Lincoln, that longest and leanest and perhaps greatest of presidents of the United States. Most people, noting the resemblance, mocked her. She, however, was proud of it, pointed it out herself, and said it went beyond the merely physical. Like Lincoln, she said, she was not afraid to make the hard decision, to stand alone against the mob. Like Lincoln, she would not complain when the odds were against her. And like Lincoln, she wanted to abolish the institution of slavery.

But she did not define the term as Lincoln did. Rather, she meant it to include the emancipation of people of all races from bondage to tobacco in general and to cigarettes, which she called "little white slavers," specifically. She might have been the most bitter and dedicated foe of the weed in the English-speaking world since James I of England. And she devoted even more of her time and energy to the cause.

Lucy Page Gaston was born in Delaware, Ohio, in 1860 and came by her reformer's zeal naturally. Her parents, motivated by their religious beliefs, were active in the abolition and temperance movements, and Gaston started teaching Sunday school at the age of thirteen. After the family moved to a suburb of Chicago, she joined her mother as a member of the Woman's Christian Temperance Union (WCTU), that euphemistically named association of females which, in truth, preferred that people abstain totally from alcoholic beverages, not use them temperately. Gaston became a writer for the WCTU newspaper, the *Union Signal*, and did not at first pay much attention to tobacco. Insofar as she thought of it at all, she dismissed it as more of a ghastly habit than a pressing issue. No one in her family smoked, nor did anyone with whom the Gastons normally associated.

But the WCTU's Department of Health and Hygiene had been looking into the effects of smoking and was becoming alarmed, finding that

> as liquor impaired the moral sense, so nicotine impaired the capacity to love and thus contributed to broken marriages. Nicotine could also inflict instantaneous physical injury, a Health and Hygiene pamphlet reported and, to prove it, cited the case of a boy whose face was totally disfigured after he had smoked a few cigarettes, and of another, a fourteen-year-old, who dropped dead immediately after his first cigarette.

To the WCTU, tobacco was a problem both in its own right and because it led to alcohol, which, in turn, "leads to the devil." Frances Willard, the WCTU's guiding light and the daughter of a snuff-sniffing mother, believed that "no man would ever be seen with a woman who had the faintest taint or tinge of tobacco about her ... it isn't thinkable." As a result, the group not only required its members to abstain from the weed, which they were without exception willing to do; it urged them to collect all the cigar and cigarette butts they saw on the streets, and then to tear them to bits and throw them away before desperate smokers could gather them up and relight them.

In time, Gaston began to concentrate more on tobacco, less on hooch. Perhaps, as she grew older, her interests flowed in that direction of their own accord; perhaps, as the journalist Frances Warfield suggests, there was something more calculated about it. Perhaps Gaston thought that "temperance work was already too well generaled. A complete, high-pressure, Heaven-guided organization in herself, Miss Gaston could not then, nor could she ever, endure working within organizations. She saw the anti-cigarette front relatively unguarded ... and made for it pell-mell." In other words, Warfield speculates, Gaston was not so much a tobacco reformer as a human being reformer. The leaf, in Warfield's view, might simply have been her medium of choice.

In time, Lucy Page Gaston would be to smokers what the more famous Carry Nation was to drinkers: a conscience, a scourge, a tireless enemy powered by unshakable conviction, and, despite it all, to her great frustration, a bit of a joke. Gaston was the more dignified of the two, never swinging a hatchet for her cause as Nation did in a number of Midwestern saloons, never screaming at miscreants on the streets, seldom losing her composure in any setting. Nor did she seek publicity the way Nation did, selling autographed photos of herself and souvenir hatchets later in life. For Gaston, it was results that mattered, not column inches or net proceeds, and she was never comfortable on center stage.

But Gaston admired her temperance counterpart, thought her a fine example for the young, and always spoke glowingly of the use to which Nation put both her blade and her fame. And the respect was mutual. In her autobiography, Nation writes, "Oh, the deadly cigarette. Thank God for the work of Miss Lucy Page Gaston of Chicago."

Their styles, then, were different, but a fire of the same temperature, ignited by the same kind of zeal, burned within each—as did a tendency toward incendiary language. When President William McKinley died of an assassin's bullet, Nation told a reporter that he probably would have survived "had not his blood been poisoned by nicotine."

History records but a single meeting between the two women, its date uncertain. It seems that Nation walked into Gaston's office one day and was stunned to see a picture of McKinley's successor, Theodore Roosevelt, on the wall. "With phenomenal restraint," Frances Warfield writes, "Mrs. Nation requested permission to tear the picture up. She had it from three eye-witnesses that the President was a cigarette smoker."

Gaston told Nation that, as far as she knew, tobacco was a stranger to Roosevelt's lips. He was, after all, the principal advocate of the strenuous life, the American embodiment of clean living and tireless effort, and how could a person promote such attitudes and at the same time deplete himself by inhaling tobacco?

Nation shook her head. Roosevelt did not practice what he preached; like so many other politicians, he was a fraud.

Gaston insisted he was not.

Nation offered a bet.

Gaston took it.

It is not clear what happened next. According to one account, the ladies wrote to the White House and put the matter directly, Gaston signing the letter in her usual manner, "Yours for the extermination of the cigarette." By return mail came a note assuring the two reformers that the president of the United States was not a smoker, a chewer, or a sniffer. He would, on occasion, accept a box of cigars as a gift, but only to be polite, then he would make a present of it to someone else. (Or, as Roosevelt himself said in his *Autobiography*, he received a single cigar from "one of my prize-fighting friends" and put it in his pocket.) The letter was signed by the president's secretary, William Loeb.

Nation was dubious, believing that Roosevelt had now taken his fraudulence to the extreme, lying in the most bald-faced of fashions to *her*. But she could not prove it and had no choice but to give Gaston the money and lament the fact that Theodore Roosevelt's picture remained in place on her friend's wall.

But there was no picture of the president's daughter on Gaston's wall. There might not even have been a picture of her on a White House wall, for Alice Roosevelt was a trial to her father, as well as to reform groups of all persuasions. As Kathleen Dalton tells us, "The W.C.T.U., Christian Endeavor Societies, the Anti-Cigarette League, and even an Ohio Suffrage Club called on her as a prominent woman to set a good example by stopping smoking. [Her father] forbade her to smoke under his roof, so she climbed on top of the roof to smoke."

Perhaps because of her dismay at Alice Roosevelt's very public tobacco use, not to mention her refusal to abstain from alcohol, Nation seems to have paid a personal visit to the president. Or tried to:

> The White House guard was polite but firm. He met Mrs. Nation before she got to the door to inform her it was not possible to see the President. When she began a harangue about cigarette fiends, the guard broke in. "Madam," he said, "do not make a lecture here." Mrs. Nation sighed, and left with a well-turned phrase. "I suppose," she said, "you have the same motto here in the White House that they have in the saloons, 'All Nations Welcome But Carry.'"

Unlike Nation, Gaston never married. She was never engaged, never had a serious beau, did not even take a fling at romance except for a single relationship in her youth about which nothing is known except that Gaston was not pleased with it, apparently seeing it as a form of weakness on her part, or a distraction from nobler purposes, or both. She prayed for it to end, and it did. So when she finally broke away from the WCTU and founded the Chicago Anti-Cigarette League in 1899, the first group of its kind in the United States, she was able to devote herself to it totally.

And she needed to, for the times were against her little assembly in a number of ways. For one, new blends of tobacco were being developed with increasing frequency, combinations of the best Virginia leaf and imports from Turkey, Russia, and other countries. As a result, cigarettes kept getting milder in taste and smoother on the draw and more and more tempting to both the novice and those who had been puffing a long time. For another, thanks to ongoing refinements in the Bonsack machine and similar devices, the prices of cigarettes kept dropping, especially relative to that of other tobacco products which could not be produced with such efficiency or in such quantity.

The place was against her, too. Bohemian and Cuban cigar makers had been settling in Chicago for several decades now, establishing tight little communities and practicing their trade avidly, seeing to it that as much fine tobacco as possible made its way into the internal organs of the city's smokers. As early as 1898, these and other establishments had transformed 27 million pounds of leaf into 153,446,000 cigars and 22 million cigarettes. In years following, they brought in even greater quantities of tobacco and began devoting a larger percentage of it to the small smokes. Tobacco was in the air, and like the even more pungent scent of Chicago's slaughterhouses, people took it as a sign of good times and economic well-being.

So it was that Lucy Page Gaston and her band of crusaders were not welcomed with open arms in the place they called home. They were, in fact, either ignored or reviled by most of those they tried to reform. But they persisted, in part because of the strength of their beliefs, in part because they directed some of their efforts at other cities, in other sections of the country, where, through lectures and mailings and the occasional sympathetic newspaper article, their message was more warmly received. A mere two years after founding her Chicago organization, Gaston was able to expand it into the Anti-Cigarette League of America. At its peak, it claimed 300,000 members and chapters in Canada as well as the United States. The figure was probably exaggerated.

Unlike the WCTU, the league did not bother linking tobacco to alcohol. As far as Gaston was concerned, cigarettes were an evil unto themselves, at least of the same magnitude as beer and wine and whiskey. They "destroyed red corpuscles, robbed the body of its vitality and the mind of its keenness, and ruined big prospects of success." A boy who was "very sick and at times acts very queer" was the victim of cigarettes. So was a seventeen-year-old lad about "to be hanged for murder." So was a young woman who "jumped from a three-story window." To all of them, and others, tobacco was either the sole cause of, or a final straw leading to, unbearable misery.

But Gaston did take a cue from the WCTU, warning that those who smoked immoderately would develop "cigarette face." It was one of the worst things she could say about a person, citing this as evidence that a human being's rotting insides would become apparent in the visage. It was also a term she never bothered to define, seeming to think that anyone who saw cigarette face would recognize it. Those so afflicted, she explained, usually "took to drink, became diseased, turned to crime, and in the end died horribly."

Gaston was especially concerned that children not become victims of cigarette face. Asked once why she dressed so austerely and, in particular, why she wore no jewelry, she smiled at her questioner as if he were a straight man and she had been eager for the line. "Thousands of clear-eyed, finely developed, clean-living young Americans are my priceless jewels," she told him.

She would never have any children of her own. Apparently, she never wanted any. She cared for them, though, in a way that was clearly maternal. She published a magazine for them called *The Boy*, "which advised America's youth how to avoid the temptations of cigarette smoking, and, if they succumbed, how to cure themselves of the consequences." And she recruited them for her organization, seeming to like young people more than adults. She fed off their enthusiasm, their malleability; she basked in their innocence, her dreams for their future.

But she also felt for those young people who were not so innocent. Frances Warfield describes her as a kind of missionary:

> A gaunt, middle-aged woman with a purpose, wearing spectacles and rustling black silk, she toured Chicago's dingier thoroughfares, in search of boy smokers.
>
> Seeing one, she accosted him with the words, "You're just the boy I'm looking for." Swiftly and graphically, she informed him where he was headed. Other boys collected. Pamphlets and tracts popped from her bulging handbag. The boys were invited to think things over. Cannily, she refused to let them swear off on the spot. Once a boy had taken thought, had reported at headquarters, had memorized and signed the Clean Life Pledge, and had received a Clean Life button, he was saved. For, said Miss Gaston, "a boy is a great stickler for honor. Once he has signed our pledge, he would cut off his hand before he would break it."

Perhaps inspired by Nation, Gaston taught the occasional use of guerrilla tactics. She instructed children to find adults who were smoking in public and sneak up behind them. When the adults seemed to be looking elsewhere, or were not paying attention for some other reason, the kids were to snatch the cigarettes from their mouths and run away as fast as they could. The adults would be angry, Gaston told her troopers-in-training; they would scream and threaten, perhaps even give chase for a block or two. But the children were not to concern themselves. The loss of their cigarettes was the best thing that could ever happen to smokers, and eventually they would realize it and look upon those who had absconded with them as saviors. Gaston referred to people who smoked as "stinkers." They, in turn, began to apply the term to her kid commandos.

But she did not stop there. Gaston pleaded with merchants not to hire boys and girls who smoked or, if they did, to insist that the youngsters sign pledges to give up the weed as a condition of employment. She demanded that Chicago appoint a special force of anti-cigarette policemen who would not only arrest offenders but lecture them on the spot about the self-destructiveness of their ways. She harangued the Chicago Cubs about the example they were setting for boys and girls by their use of the leaf, both cigarette and plug. And she once stormed out of the Chicago Opera's performance of *The Secret of Suzanne* after the female lead lit up her third cigarette. When, later, a reporter for the *Tribune* asked what was the matter, Gaston sputtered at him, barely able to contain herself. "Horrible," she said. "Horrible. One after another. I saw her with my own eyes. It is enough to turn one forever against grand opera. An artful embellishment of a pernicious vice which should receive the stamp of disapproval from every true American woman."

Lucy Page Gaston had a dream. She wanted her Anti-Cigarette League to become as powerful as the Woman's Christian Temperance Union and the Anti-Saloon League, the latter of which would lead the United States into Prohibition a few years hence. It never happened. The consequences of too much drinking were so noticeable, those of too much smoking so hidden. Gaston was only the latest of her breed to be victimized by the weed's pernicious subtlety and forced to follow temperance's lead and hope to reach the same destination.

She did, however, manage to attract some notable support. The humorist Elbert Hubbard, an A-list celebrity of the time, told his readers that "cigarette smokers are men whose futures lie behind them." David Starr Jordan, the president of Stanford University and a better-known figure than most in the academe, said, "Boys who smoke cigarettes are like wormy apples that fall from the tree before they are ripe." Jordan worked zealously for the Anti-Cigarette League: holding office, making personal appearances, and writing slogans for posters.

William Booth, founder of the Salvation Army, listed "Fifty-Four Objections to Tobacco," including his beliefs that it "tends to insanity," "is powerful in leading to forgetfulness of God," and "arrests the growth of the young." To Sir George Williams, who organized the YMCA, cigarettes were a serious problem growing worse by the day. It was a view often expressed in both the group's classes and its literature.

John Wanamaker took Gaston's advice and refused to hire smokers to work in his Philadelphia department stores. Executives at railroads like the New Haven, the Rock Island, and the Atchison, Topeka, and Santa Fe did the same. If a smoker somehow got by the screening process and made his way onto the tracks, he was fired at first puff.

Most of the preceding donated money to the Anti-Cigarette League, as did Andrew Carnegie; Julius Rosenwald, president of Sears, Roebuck, and Company; and William C. Thorne, president of Montgomery Ward. Other donations came from other barons of American commerce, men who might have been less well known but were well endowed in their own rights and willing to contribute to a socially responsible cause.

Still, Gaston had a hard time making ends meet. In a typical month, it has been reported, the league spent $1,250 and took in somewhat less. Her wealthy allies, she began to grumble, offered more allegiance than wealth. She sometimes badgered them for greater contributions. She seldom succeeded, and the more she badgered, the more *they* began to grumble.

Henry Ford was a particular vexation to her. A generous contributor to the Anti-Saloon League, Ford provided much in the way of verbal support, little—possibly even nothing—in the way of financial support to Gaston. But he agreed with her that cigarettes were a curse to civilization, saying he "was convinced [that they] contained the vilest poison known to man." To get the point across, he even published a pamphlet called *The Case against the Little White Slaver*. In it, he quoted an explosives maker named Hudson Maxim, who believed that the slavers were as dangerous to mankind as his own products. "If all boys could be made to know," Maxim wrote, "that with every breath of cigarette smoke they inhale imbecility and exhale manhood, that they are tapping their arteries as surely and letting their life's blood out as truly as though their veins and arteries were severed, and that the cigarette is a maker of invalids, criminals and fools—not men—it ought to deter them some."

And in April 1914, Ford received the following letter from his friend Thomas Edison, who not only shared his sentiments but had thought the matter through in some detail:

> The injurious agent in cigarettes comes principally from the burning paper wrapper. The substance thereby formed is called "Aerdein." It has a violent action on the nerve centers, producing degeneration of the cells of the brain, which is quite rapid among boys. Unlike most narcotics this degeneration is permanent and uncontrollable. I employ no person who smokes cigarettes.

John L. Sullivan, the heavyweight boxing champion of the world, was not one for analyzing the chemical content of cigarette paper, which Edison did mistakenly. For his own reasons, though, he shared the inventor's feelings, as well as the stereotypes of a previous generation. "Smoke cigarettes?" he once said to a reporter. "Not on your tut-tut. ... You can't suck coffin-nails and be a ring champion. ... Who smokes 'em? Dudes and college stiffs—fellows who'd be wiped out by a single jab or a quick uppercut."

But Sullivan's position was not as simple as it seems. The champ, you see, was a cigar smoker. So was Edison. In fact, there were times when the two of them posed for photographers, either blowing out smoke rings or chomping on the ends of cigars they had not yet kindled, sucking in the taste of cold, processed leaf. Their chins would be upturned, their eyes narrowed as if starting or finishing a wink; they were the very pictures of jauntiness.

And that was the terrible irony of Lucy Page Gaston's vocation: the number of people who supported it not because they were anti-cigarette, but because they were pro-cigar or pro-pipe or pro-plug, and believed that if Gaston were successful in her efforts, their own particular favorites would regain a bigger share of the market and those who consumed them would be seen as more fashionable.

It was cigar makers who cheered loudest for Gaston. In fact, inspired by her fulminations, they stepped up the pace of the slanders they had begun some years earlier, reviving the tales about cigarettes being the work of Chinese lepers and "tramps and ragpickers." More effectively, they acted as flacks for Gaston's more medically based charges, seeing to it that they were widely disseminated—the more serious the charge, the greater the publicity provided free of charge by the cigar industry.

In a way, this put Gaston in the same predicament as George Bernard Shaw's Salvation Army heroine, the pacifist Major Barbara Undershaft. Both women faced the choice of either accepting help from the other side or finding their causes gravely weakened. Both chose the former. Undershaft finally took money from her father, the weapons king, to spread the word of peace and disarmament and to finance various charitable projects, and she did it with a kind of philosophical resignation. She did not ignore the origins of her cash; rather, she concentrated on the good it could do in its new incarnation.

Gaston, however, railed at the hypocrisy she was forced to endure. She fumed, foamed, went back yet again to the Wanamakers and the Carnegies and the rest of them, imploring, almost threatening, but still failing to persuade them to up their antes. So endure the support of her foes she did and, as a result, the Anti-Cigarette League kept on as it had been, succeeding to a degree but making no big ripples.

Looked at carefully, though, as Gaston did, it was the wrong kind of success. The league seems not to have persuaded Americans of the cigarette's potential for harm so much as to have tapped into the reserves of prejudice that, in some quarters, still existed against it, encouraging people to think that the small smoke was as unmanly as ever, despite the rugged new names and bargain-basement prices. Sullivan put it this way: "It's the Dutchmen, Italians, Russians, Turks and Egyptians who smoke cigarettes, and they're no good anyhow."

If they were thought to call a man's ethnicity into question, cigarettes were also thought to give a woman an air of either masculinity or permissiveness and a child the taint of unruly behavior. The impressions were solidified in the popular arts; no turn-of-the-century illustrator who drew a bum, a whore, or a delinquent finished his sketch without dangling a slaver from his subject's lips. And, parenthetically, few were the authors of dime novels who did not attach a cigar to their heroes.

Less back-handed support for the Gastonites came from various state legislatures, which were finally responding to all the time the league had spent hectoring them, inundating them with pamphlets and pleas and statistics. Between 1895 and 1909, with cigarettes already having been outlawed in twenty-one states and territories for children, twelve states made them illegal for adults. They included Iowa, North Dakota, South Dakota, Indiana, Minnesota, Kansas, Nebraska, Oklahoma, and Wisconsin, the same heartland venues that were so receptive to the temperance movement at the same time.

In New Hampshire, it was "illegal for any person, firm, or corporation to make, sell, or keep for sale any form of cigarette." Legislators in Illinois decreed that "the manufacture, sale, or gift of a cigarette 'containing any substance deleterious to health, including tobacco,'" would result in a fine of $100 or a jail term of a month for the guilty party.

But the states did not always enforce their laws. The Wisconsin ban, for example, was "flagrantly and openly violated." In Washington, sixty-six arrests were reported in the first four months of the cigarette proscription but not in a single one after that, and after a few years, the law was declared unconstitutional. The *New York Times* applauded:

> The smoking of cigarettes may be objectionable as are many other foolish practices, and it may be more injurious than other modes of smoking tobacco, but it is an evil which cannot be remedied by law, and it is not the kind of evil to the community at large that is a legitimate subject for legislative action. That kind of a law is pretty sure to be evaded, and it begets a contempt for law in general and for public authority that is more pernicious than selling cigarettes or even smoking them.

But it was not just a dislike of the small smoke that led to the legislation, however ineffective it was proving to be; another factor was a dislike of those who most avidly consumed it. Many Midwesterners thought of cigarettes, like alcoholic beverages, as playthings of the swarming immigrant masses that infested the big cities of the East. They were "inferior breeds of people," these citizens of the heartland believed, Catholics and Jews while they were Protestants; Southern and Eastern Europeans while they had their roots in American soil, if in some cases no more deeply than a single generation. The immigrants were frightened and hopeful and edgy about their prospects in their new home. the Midwesterners had gotten over all that and did not need cigarettes to work off their tensions or placate themselves for some other reason. In voting against them, then, a good number of people were voting their xenophobia, their longing for a nation like the one they had known, or imagined, before.

In 1921, two more states prohibited cigarettes, bringing the total to fourteen in the past quarter-century. That same year, ninety-two bills to ban or limit the sale or manufacture of small smokes were under consideration in twenty-eight other states. Some would pass, some would not, but taken together the measures were a genuine threat to the tobacco industry, the first it had ever faced.

The industry had no choice but to respond and did so by adopting some of the same high-pressure lobbying techniques as the Anti-Cigarette League. Then it went further. On one occasion, James Duke sent some minions to Chicago, Gaston's own front yard, to offer an alderman a $25,000 bribe to vote down an anti-smoking ordinance. On another, he informed legislators in Tennessee that $500 awaited each of them who would turn his back on the reformers' pleas. In neither case was Duke successful, but he and other tobacco executives kept their eyes and wallets open in these years, constantly surveying deliberative bodies at the state and local levels, constantly ready to pounce if it seemed a vote might go against them.

But there were so many bodies, and not enough money, even in the tobacco coffers, to subvert them all; there were, as a result, far more laws than successful pounces. Some of the legislation passed in this period restricted smoking according to age, some according to gender, some according to the location of the deed, with New York City banning tobacco for a time on trains, trolleys, and ferries. Penalties varied; a fine of twenty-five dollars was typical. And, as was the case at the state level, the standards of enforcement varied. A cop's mood on a given day, or his own particular cravings or lack thereof for tobacco, might be all that determined a smoker's fate.

To the Quakers, cigarettes were an "evil of great magnitude," and officials of the church urged their brethren not only to refrain from them on their own but "to labor in a spirit of love with their members who use or grow tobacco." At the Quakers' general conference in 1900, tobacco in all forms was prohibited. Previously, the Methodist Episcopal church of the North had determined that preachers seeking admission to the faith had to answer yes to the question, "Will you wholly abstain from the use of tobacco?" And before that—in fact, in the year of Lucy Page Gaston's birth—the same group's General Conference in Ohio had resolved that, "after the present session, we will not receive any person into full communion who persists in the use of tobacco."

A few other denominations, and a few individual churches here and there, also demanded that their ministers and congregants swear off the leaf. For the most part, though, America's clergy were silent on the issue of the little white slaver. Despite Gaston's own religious convictions, strong enough to have motivated her to teach Sunday school at so young an age, her movement began as, and remained, primarily a secular one.

But the lack of assistance from the pulpit does not seem to have troubled her. Nor did she go to any lengths to change the clerical mind. What she did do, in addition to pursuing the backing of lawmakers and her corporate donors, was seek endorsement from the medical community, and it gradually began to come. A number of doctors aligned themselves with the Anti-Cigarette League, some for honorable reasons, others for a chance to make a fast buck. A few from each camp opened clinics in New York and Chicago and other large cities, storefront establishments and holes in the wall, new and sometimes dubious enterprises in old and rundown neighborhoods. They promised to help their patients quit smoking and never even strike a match again, much less fire up and inhale a cigarette. But in the main, what they provided were placebos and quack remedies, platitudes and simplistic literature. There is almost no information available on the cure rates of such places and, therefore, little reason to believe that they were impressive.

Among those who got in on the action, opening a clinic of his own, was the secretary general of the Anti-Cigarette League, Dr. D. H. Kress. As Robert Sobel tells us, Kress "patented a mouthwash that contained a weak solution of silver nitrate. Gargle with it after every meal for three days, said Dr. Kress, take warm baths and switch to a bland diet. By the end of the third day your craving for cigarettes will be over." This was the clinic's regimen. To some it sounded possible, to others so tempting as to be irresistible:

> Messenger boys, chorus girls, housewives, an occasional businessman, and the idly curious trooped to the clinic—along with reporters looking for feature stories. The clinic was so successful that the league soon established a second one in Chicago, for women only, followed by others in Detroit, Cincinnati, and elsewhere in the Midwest.

Did the treatment work? Yes, at least for those firmly resolved to give up tobacco before undertaking it and thus halfway cured before their first onerous mouthful of silver nitrate. Did it "produce extreme nausea," as one critic charged? Also yes, but no one ever said the weed could be vanquished with ease.

Other nostrums were equally accessible, not only in clinics and on the shelves of pharmacies and a variety of other stores, but through ads appearing on the back pages of newspapers and magazines, ads that call to mind the claims that appear today in supermarket tabloids for breast-enlargement creams and cellulite evaporators. There were pills and potions, food supplements, and mechanical devices. Amazing, said the copy in thick black ink, astonishing, scientific, miraculous—you won't believe the results! One of the more successful products—for the company that sold it, if not for the smoker who wanted to quit—was something called No-To-Bac, "the only guaranteed, harmless, economical cure for the Tobacco Habits in the world." Less than ten cents a day.

More reputably, the Anti-Cigarette League advised people to eat healthful meals, get plenty of rest and exercise, and chew a gentian root at those moments when temptation for the weed was most acute. Gaston always carried one in her purse. "Remember gentian root," she would tell people who were struggling with tobacco, or at least those who wondered about that strange looking thing she had just pulled out of her bag. "You spell it g-e-n-t-i-a-n."

It was a remarkable set of developments. For so many years, tobacco had been a thriving industry, a cornerstone of the nation's economy going back to the days when it was the entirety of the Jamestown economy. Now, although not exactly thriving, there was a new American business whose specific goal was to cut into the business of the older business. Cigarettes and smoking cures—those who provided them were not competitors jostling for market share; they were adversaries fighting for existence. It would be, until several more decades had gone by, the most one-sided of struggles.

AS IS THE CASE with reform movements of virtually all kinds, this one led true believers to occasional outbreaks of outrageous, and sometimes comical, excess. Early in the twentieth century, for instance, a traveling opera company staged Bizet's *Carmen* in a series of small towns in Kansas. Afraid of riling the inhabitants, the producers set the opera on a dairy farm instead of a cigarette factory. When the title character made her initial entrance, she carried a pail of milk, not a sheaf of tobacco leaves. And instead of references to rolling tobacco into precisely cut sheets of paper, there were references to squeezing udders and filling up buckets.

In addition, a number of schools throughout the country agonized over, and in some cases actually banned, the nursery rhyme "Old King Cole." Parents tore the offending page out of their children's books. Clergymen applauded their actions. His majesty, after all, had had the effrontery to call for his pipe!

More traditionally, reformers would sometimes picket stores that sold tobacco products, alternately praying for the people who bought them and damning them to hell. They would hand out brochures, sing songs, sometimes cough in the faces of smokers to show them what was ahead. They were like the Women's Crusaders, the foes of alcohol who had preceded the WCTU, stationing themselves in front of taverns and, if nothing else, embarrassing tipplers with the cold and unrelenting piety of their commitment.

Viewed from the present, it seems at least a little puzzling—so much opposition to tobacco, so many laws and cures, the bowdlerizing of texts, and the confrontations and public prayer sessions. It cannot explained by the weight of evidence against the weed, for there was little at the time, and there would not be substantially more until two things occurred. First, the X-ray and other techniques of internal medicine would have to be developed or improved, and the people employing them would need to understand precisely what they were looking at and how to track down its causes. Second, more sophisticated methods of accumulating and analyzing data would need to be conceived, so that an individual case could be understood in its larger and more accurate context.

But in the meantime, something was happening beneath the surface of events and attitudes—something formative, important, complex. A kind of groundwork was being laid for later discoveries, a foundation upon which the eventual case against tobacco would be erected. Americans still could not prove that smoking was dangerous, but they were moving closer to this knowledge; they had, in fact, set foot on the very next rung of the ladder. They had begun to believe that the weed was unclean and to worry about the implications.

For most of our nation's history, this would not have mattered. Wallowing in the muck was an inescapable part of opening up the wilderness, pushing back frontiers, and establishing a new society. In colonial times, "ladies and gentlemen of high morality and august standing could ignore the way everybody smelled because, since everybody did, it made no given individual objectionable." Most Americans remained pungent, to one extent or another, for more than a century thereafter, and because no solution readily presented itself, no cultural opprobrium was attached.

How could it be? Stench was everywhere. Human beings and animals both produced it, and neither group was very good at cleaning up after itself:

> It is difficult for those reared after the automobile ousted the horse to realize how excrement thus pervaded the outdoors of the nineteenth-century city, making it a sort of equine latrine. ... [The dung] dried in the summer sun to become high-flavored dust blowing into mouth and lungs. In wet weather it incorporated into the mud that was tracked everywhere, finding the bottoms of women pedestrians' long skirts, gradually impregnating the straw on the floor of the omnibus until the passengers might as well have sat with their feet in a horse stall needing cleaning.

The dust and mud drifted and oozed into water supplies, as did the quantities of waste that were deposited directly into rivers, streams, and lakes. Taking a bath was not a popular or convenient activity anyhow, but under these conditions it was worse than a low priority; it was a fearful prospect. A person could get cleaner and sicker in the same instant. In fact, many Americans felt about a bath as had their European forebears in the time of the plague: that to remove dirt from the pores was to open them to various kinds of ailments that were lurking in the breeze. Dirt, in other words, was thought to be a sanitary coating for the epidermis.

In Philadelphia late in the nineteenth century, the city council considered a law to prohibit baths in wintertime. It almost passed. The councilmen's counterparts in Boston did pass a bill making baths illegal unless they had been specifically recommended by a doctor, and few were the men of medicine who would do such a thing, doctors being among the leading non-bathers of the period.

In Cincinnati, it was not just the domestic manners of the Americans that troubled Fanny Trollope. So, too, did the lack of clean water, which led in the summer to the rampant breeding of mosquitoes. Trollope and one of her sons came down with malaria during their stay. To another son in London she wrote, "You would never know Henry and me, we are both so thin." Had she visited Cincinnati several decades later, she could just as easily have lost a similar amount of weight.

As the nineteenth century drew to a close, though, scientists were learning the perils of polluted water and passing the information along to people who became slowly but surely horrified by it and eager to take action. In New Orleans, some well-to-do ladies formed themselves into the Women's League for Sewerage and Drainage—an unglamorous name, but an important mission. The group forced a special election on a proposal to spend tax money for a new sanitation system. Never before had a project of this sort been proposed in New Orleans, or almost anywhere else in the United States, and the women campaigned tirelessly on the measure's behalf. When it passed, they celebrated not only a cleaner city but their own nascent political power.

The story was the same in Chicago. To prepare for the World's Fair of 1893, writes Erik Larson, "The city stepped up its efforts to remove garbage and began repaving alleys and streets. It deployed smoke inspectors to enforce a new antismoke ordinance. Newspapers launched crusades against pestilent alleys and excess smoke and identified the worst offenders in print."

Had Benjamin Rush been around, he would have shaken his head in wonder. He had known about the importance of public hygiene several generations earlier. What had taken his fellow Americans so long to catch up? Why, he might have asked, is a voice raised ahead of its time so seldom heard?

Similar movements began elsewhere in the United States. Drainage systems were built or improved in numerous cities; trenches were dug to carry away sewage and were monitored so they would not overflow. In some cities, men were employed as street disinfectors, a uniformed force that patrolled the thoroughfares with mops and buckets and constantly wrinkled noses. They did not clean up all the fecal matter, urine, and rubbish; it would have been too big a job for twice their number. But they did their best to render the foul materials harmless with an array of chemicals and to destroy the insects and rats and other bearers of pestilence that hovered in their midst.

Drinking water soon became filtered. Milk was pasteurized. The manufacturers of other beverages mopped the floors of their workplaces and hosed down the machinery after every couple of shifts. A new saying entered the language, its origin uncertain. "Cleanliness is next to Godliness," it went, and quite a few people believed it.

At about the same time, researchers such as the German Robert Koch were taking advantage of powerful new microscopes, spying on previously invisible organisms and positing the germ theory of disease. Unlike many scientific advances, which are too arcane to be understood by the layman, this one was big news and easily comprehensible, at least in its basics. It was, therefore, a perfect story for the American tabloid press, then at its yellowest and most sensational: "The Sunday papers had horrendous pictures of flies enlarged to the size of frogs," it was observed, "all fearsome eyes and bristly legs, and texts explaining how diligently they carried germs from privy to table. Hence 'Swat the Fly' campaigns and widespread—and long overdue—installations of window screens and screen doors."

The U.S. Army developed incinerators and encouraged municipalities to do the same. Large, public garbage dumps became less common and began to disappear altogether from major population centers. The casual disposal of waste in public places was met with either frowns or fines. City streets began to take on entirely different aspects and odors.

And it was all because people were finding out about dirt, the whole story: how it could make a cut or wound worse by infecting it, and how untreated cuts and wounds could lead to illness and worse. They were also finding out about other kinds of maladies that could be caused by the neglect of hygiene in kitchens, restaurants, meat-packing houses, canneries, schools—anywhere that food or beverage was handled.

Or where human beings were handled. Some time earlier, Dr. Oliver Wendell Holmes, the father of the noted jurist, declared that puerperal fever, which took the lives of many women after they had given life to their own children, was often spread by the uncleanliness of the doctors who treated them. It was a startling claim for the time, and Holmes's fellow physicians resisted it as long as they could.

As the years went by, men of medicine realized that other diseases could also be transported through a lack of proper emphasis on cleanliness. So they washed their hands, brushed their teeth, sterilized their instruments, and even began to take baths, summer *and* winter, and to encourage their patients to do the same.

Primitive as it was, the technology of the time contributed to the newfound trendiness of hygiene. It made personal and civic cleanliness easier to achieve and, in a few cases, more fun. Some communities built public urinals. Others built public lavatories and provided disposable paper covers for the seats. Private homeowners who could afford to knock down their outhouses replaced them with indoor bathrooms that had running water and flush toilets. Those less affluent might collect their wastes in large, rubber-sealed containers and see to it that they were properly emptied at least once a week. A person could buy a "copper" for his home and heat water in it, boiling away impurities; he could spend more money and buy a hot-water boiler; he could rob a bank and purchase an expensive apparatus called a Velo-Douche, a cross between a stationary bicycle and a shower, with the peddling motion of the legs pumping and recycling the water.

Cities, states, and the federal government set up health regulatory agencies. In the wake of *The Jungle*, Upton Sinclair's novelistic expose of the meatpacking industry, Congress approved legislation to make beef on the hoof safer. It decreed that navigable waters be less polluted, and it allocated money for several other sanitation-related issues, the first time it had ever done such a thing. Americans were becoming exposed to the truth, and the truth was making them healthy.

In an atmosphere like this, the cigarette could not help but fall under suspicion. It smelled bad; it tasted hot and acrid; the smoke reminded some people of the fumes that poured out of factory smokestacks, and these too, it had been shown, were a form of pollution. Cigarettes just plain seemed dirty, and dirty, men and women now knew, mattered. In fact, as early as 1892, the U.S. Senate Committee on Epidemic Diseases called cigarettes a public-health hazard, although it was not able to cite hard evidence, and few people took the charge seriously.

As a result of all this, cigarette sales declined in the United States between 1897 and 1901, the only drop ever recorded up to that time. The reformers and researchers had something to do with it, of course, but so did the federal government, as it raised taxes on the weed to finance the Spanish–American

War. The United States won. It should have. According to a member of the British Parliament, who was as hostile toward the American foe as any of his predecessors of centuries past, Spain could not possibly have prevailed. The country was riddled with tobacco, had been for a long time; its fighting men did not have a chance.

However, sales of the small smoke began to increase as soon as the war ended. They would continue to rise, steadily if not spectacularly, for the next decade and a half.

Then the country went to war again, this time on a grander scale than ever before. More nations were involved than there had been in Cuba, more soldiers, more territory. The stakes were much higher now for a far greater number of people and causes. It was the beginning of the end for Lucy Page Gaston.

NO SOONER had the United States taken up arms against the Central Powers in 1917 than General John J. Pershing echoed the sentiments of George Washington a century and a half earlier. "You ask me what we need to win this war?" Pershing said with a rhetorical flourish. "I answer tobacco as much as bullets." Later, from his command post with the American Expeditionary Force in France, the general got more specific. In a cable to his superiors in Washington, he demanded tens of thousands of tons of cigarettes, the quick and portable smoke, so easily adaptable to the conditions of battle. It was too much, a wildly impractical amount. Pershing got it, and then some.

"A cigarette may make the difference between a hero and a shirker," said the general's top aide, Major Grayson M. P. Murphy, and among those who believed him was President Woodrow Wilson, who announced his support for the *New York Sun*'s "Smokes for Soldiers Fund." He urged citizens to donate, urged soldiers to partake if so inclined, urged them to *be* inclined.

Also taking the smokers' side were a large number of doctors, men who either did not agree that the leaf was unclean or else, although sharing the reformers' views, had decided to keep their objections to themselves for the time being. Several army physicians put their endorsements of tobacco on the record in both speeches and newspaper interviews. A navy physician declared that cigarettes were "a means of diversion which, far from interfering with a man's performance of duty, attaches him to it and renders it less burdensome." Other doctors, though perhaps guided more by superstition or wishful thinking than medical acumen, thought that cigarettes might vaccinate people against influenza. The conflict in Europe, it seemed, was going to be even more of a boon to the tobacco industry than the financial panics of the previous century:

"Coffin nails" was what we said
But the war has changed the name.
The cigarette is now first aid
In this hellish, killing game.

Celebrities promoted both the cause and themselves. Performing on various bills to raise money for combat-bound cigarettes were Will Rogers, Fanny Brice, W. C. Fields, Lillian Russell, Eddie Cantor, and Ethel Barrymore, among others. They sang songs, told jokes, did some fancy stepping, urged their audiences to contribute whatever they could. "Sophie Tucker," it was reported at the time, "appeared at so many benefits that she was called 'The Smoke Angel.'"

Lucy Page Gaston was not pleased by the renewed perception of tobacco as friend of the man in need. But she was absolutely livid at the volunteer organizations that were distributing the leaf overseas. One of them was the YMCA, which did a complete about-face, not only withdrawing its support from the Anti-Cigarette League but publicly criticizing the group for refusing to do an about-face of its own in the present circumstances. A former member of the league who signed on with the YMCA abroad explained his conversion as follows:

> There are hundreds of thousands of men in the trenches who would go mad, or at least become so nervously inefficient as to be useless, if tobacco were denied them. Without it they would surely turn to worse things. Many a sorely wounded lad has died with a cigarette in his mouth, whose dying was less bitter because of the "poison pill." The argument that tobacco may shorten the life five or ten years, and that it dulls the brain in the meantime, seems a little out of place in the trench where men stand in frozen blood and water and wait for death.

The YMCA shipped more than 12 million dollars' worth of tobacco products to European battlefields in America's year and a half at war. When the going got tough, the Y got dogs, strapping cartons of cigarettes to the backs of all manner of canines, large breeds and small, to transport them quickly and safely to the front lines.

The Salvation Army also turned on Gaston, no longer expressing fifty-four objections to the leaf, now seeming to have as many reasons, if not even more, to endorse it. Booth's group, too, forwarded cigarettes to the front lines and "actually encouraged soldiers to smoke them in the interest of chastity and sobriety. Men who were offered the comfort of a cigarette, it was argued, would be less likely to seek more harmful diversions."

The Red Cross, too, was offering comfort—nicotine-laced and packed tightly into white-paper tubes. One soldier claimed that Red Cross nurses were "one of the greatest blessings on earth," as they not only dispensed cigarettes but lit them as a special favor to men with bandaged hands, broken arms, and shattered morale.

The volunteer groups collected the cigarettes from various drop-off points in the United States, then loaded them onto trains that rolled across plains and prairies, foothills and deltas, farmlands and urban landscapes, chugging their way to one port or another, where ships awaited with yawning cargo holds. Sometimes banners were hung from the cars, and if he had the proper vantage point, a person could read them from a mile or more away.

AMERICA'S BEST
FOR AMERICA'S BRAVEST

SMOKE OUT THE KAISER

WHEN OUR BOYS LIGHT UP
THE HUNS WILL LIGHT OUT

Uncle Sam did his part, too, waiving tax and export restrictions on tobacco products and seeing to it that the needs of the volunteer organizations, for both manpower and logistical support, were met as promptly and efficiently as possible. "Any man in uniform in 1917–18," it has accurately been observed, "in almost any part of the United States or France, could be certain of finding a canteen where he could get free coffee, doughnuts and cigarettes."

It was the same elsewhere. Iain Gately writes that "the British infantryman's tobacco ration was 2 ounces per week, while the German's daily ration was two cigars and two cigarettes or 9/10 oz. pipe tobacco, or 1/5 oz. plug tobacco, or 1/5 oz. snuff, at the discretion of his commanding officer. Britain's sailors and marines did better than the foot soldiers: the Royal Navy tobacco ration was 4 ounces per week."

Some companies donated their goods. Others sold them to the distributing agencies at cost or less, and the agencies solicited donations to pay for them. Either way, it was an expensive proposition for the leaf industry, especially the cigarette makers, and while the war continued they lost money with every pack, every shipment. They said they did not mind. They were doing their bit for the boys.

But they were also doing a fair amount for themselves, believing that the combination of combat-produced anxieties and free cigarettes to ease them would eventually work to their advantage. It would make casual smokers out of non-smokers, heavy smokers out of casual smokers, and nicotine-gulping, fume-spewing addicts out of heavy smokers. Once the fighting was over, profits would skyrocket, as the machines that had descended from the original Bonsacks would keep on working twenty-four hours a day, seven days a week, churning out product to meet ever greater demands—and this time, the product would sell at full retail price. The cigarette firms would make more than enough money to cover their previous losses. In other words, they looked at World War I as, among other things, a gigantic loss leader.

And a fellow named William D. Parkinson, writing in the *Boston Herald* and perhaps smoking something more mind-altering than mere leaf, heartily approved of the situation, finding "something almost inspiring in the spectacle of a great industry capitalizing [on] war as did the tobacco industry. Every recruiting poster [showing a manly figure with a cigarette] was a gratuitous advertisement for the cigarette."

Magazine ads made the same point. One, in particular, featured a drawing of a doughboy whose uniform was wrinkled, his face dirty, and his eyes leaden with fatigue. "But the helmet is at a cocky angle, signifying spirit and resolve," writes Robert Sobel. "There is a knowing smile on his face as if to say the American can take anything thrown at him, and then return for more. The reason is also clear; between his lips is a freshly lit cigarette." But not just any cigarette. The brand, in this ad, is a Murad, the same name as the Turkish sultan of long ago who had executed as many as eighteen smokers in a day. Centuries after his own death, the villain had been co-opted by the other side.

When the war ended, so, once and for all, did the small smoke's image problems in the United States. If a cigarette was good enough for men who were being assaulted by grenades and mortars, Big Berthas and howitzers, good enough for men who were standing their ground against tanks and keeping midnight watch aboard ship for the sneaky menace of submarines, good enough for men who were slogging their way through muddy fields with landmines all around them and clouds of mustard gas wafting behind, good enough for men who, unlike soldiers in any war before them, were being fired on from the skies, of all places, the very heavens raining down destruction on them—if a cigarette was good enough for men like these, men who were going through the nine circles of hell for the sake of democracy, then they were good enough for everyone else, every single American of every single type. Cigarettes got a soldier through his days and nights; they were a pleasure where it did not seem that pleasure could exist, even for a few moments.

"Conversely," writes Richard Klein of the world of fiction, "the worst moments in war are frequently represented by a character hating the taste of a cigarette." Here is William Styron's character in the novel of Korea, *The Long March*, at a moment of desperate exhaustion: "Blood was knocking angrily at his temples, behind his eyes, and he was thirsty enough to drink, with a greedy recklessness, nearly a third of his canteen. He lit a cigarette; it tasted foul and metallic and he flipped it away."

Floyd Gibbons, a real-life marine bloodied in combat, tells of his experience in a field hospital on the western front:

> My chest was splashed with red from the two body wounds. Such was my entrance. I must have looked somewhat gruesome, because I happened to catch an involuntary shudder as it passed over the face of one of my observers among the walking wounded and I heard him remark to the man next to him:
>
> "My God, look what they're bringing in."
>
> Hartzell placed me on a stretcher on the floor and went for water, which I sorely needed. I heard someone stop beside my stretcher and bend over me, while a kindly voice said:
>
> "Would you like a cigarette, old man?"
>
> "Yes," I replied. He lighted one in his own lips and placed it in my mouth. I wanted to know my benefactor. I asked him for his name and organization.
>
> "I am not a soldier," he said. ... "My name is Slater and I'm from the Y.M.C.A."
>
> That cigarette tasted mighty good. If you who read this are one of those whose contributions to the Y.M.C.A. made that distribution possible, I wish to herewith express to you my gratefulness and the gratefulness of the other men who enjoyed your generosity that night.

But there were people who had recognized the importance of the weed in the Great War even before the United States took up arms:

> Americans who followed the war in the newspapers during the neutrality phase in 1914–17 would see pictures of English, French, German, and Russian soldiers at rest in the backlines, more often than not smoking cigarettes. ... [I]mportant French and Belgian politicians trekked to the front lines to be photographed handing out packs of cigarettes to men coming off the line. Americans started The Belgian Soldiers' Tobacco Fund, to raise money for cigarettes that were distributed gratis in that army. This transpired after a news story that some Belgians sent an appeal to their minister of war. "Give us worse food if you like, but let us have tobacco."

These were no longer foreigners, dandies, and snobs whose habits were to be avoided at all costs by red-blooded citizens of the United States of America. These were our brothers in the cause of freedom, our future allies. Most Americans were on their side from the start. When it became *our* side as well, we, too, would have sacrificed food for the sake of our own supplies of tobacco.

LUCY PAGE GASTON did not understand what was happening. Nor did she accept it. She fought as passionately for her beliefs during this time as the American troops fought for theirs. She wrote to Secretary of War Newton D. Baker, trying to win him over. "People seem to be entirely swept off their feet," she complained, "and the general impression prevails that as soon as a man puts on the uniform he must begin to dope up preparatory to a possible trench experience. This, of course, is the greatest folly."

She repeated her old warnings about the destructive effects of cigarettes, the whole arsenal of them, issued and reissued them time and again to all who would listen and many who would have preferred not to. Then she added a new warning: Gaston claimed that the little white slavers created even more of a problem in wartime than they did in peace because the glow from a cigarette on the battlefield gave away a soldier's position. Thus, he was likely to die of enemy fire long before the toxins in the leaf did him in.

It was not an idle point. In Erich Maria Remarque's classic novel of World War I, *All Quiet on the Western Front,* a book in which tobacco products appear so often that they might as well be characters and in which the narrator finds "ten cigars, twenty cigarettes, and two quids of chew per man" a "decent" daily portion, the soldiers are given an order as night envelopes them. "Cigarettes and pipes out," they are told. "We are getting near the line."

Once again, to her immense frustration, Gaston found a segment of the tobacco industry embracing her position, even quoting her directly in some of its materials. The glow from cigarettes was worse than a telltale sign, said the chewing-tobacco interests, having floundered now for decades and looking for any way they could to drum up business. The cigarette was a beacon, a veritable spotlight, all the more so when several doughboys were lighting up in the same trench at the same time. The fine American plug makers saw it as their patriotic duty to recommend their own product instead. Nobody can see you when you're chewing, they explained. Of course, there would be all that juice to spit away, but it could be done quietly enough so that the foe would not hear. As for the mess, those puddles of leaf spittle that would be underfoot in the foxholes, well, war was hardly a tidy undertaking to begin with.

Gaston ignored the plug makers. She ignored those who manufactured cigars and pipes and pipe blends. In fact, she was soon ignoring her own constituents, even members of her own organization, fellow officers, some of whom thought she was going too far in condemning the wartime use of cigarettes. These were extenuating circumstances, a number of reformers believed. They had not changed their minds about the evils of tobacco, but neither did they think they should be expressing themselves as freely now as they had in peacetime.

Gaston, however, felt so strongly about the weed and the special treatment it was receiving that she threatened to take the matter to court, to bring suit against the YMCA and the Red Cross and all the rest of the turncoat organizations, claiming it was illegal for them to be shipping cigarettes overseas because of anti-smoking statutes on the books in states from which the shipments originated or through which they passed. The courts paid no attention to her. Unfortunately for Gaston, others did.

Up to this point, the Anti-Cigarette League of America had been perceived by many people as a relatively benign group. Those who belonged were mothers, grandmothers, salt-of-the earthers; they went to church, paid their taxes on time, attended all the civic functions. Many who did not belong to the league favored its goals—or, at least, a modified version of them—and those who opposed the league were tolerant, sometimes even bemused, willing to credit it with having its heart in the right place. At worst, the league was perceived as little more than a nuisance, a butt of jokes for sophisticates but hardly a serious threat to pleasures and values of long standing.

Not any longer. Now there were people who thought that Gaston, in her intransigence, had actually become a traitor. A few went so far as to wonder whether the Espionage Act of 1917 could be interpreted in such a way that she was in violation. In part, the act reads as follows:

> Whoever, when the United States is at war, shall willfully make or convey false reports or false statements with intent to interfere with the operation or success of the military or naval forces of the United States ... shall be punished by a fine of not more than $10,000 or imprisonment for not more than twenty years.

The wording was not relevant. The act could not be used against Lucy Page Gaston. Nonetheless, many Americans, non-smokers included, believed that the lady had done the unforgivable, putting her own interests, however well intentioned, above those of the nation. Right cause, perhaps, but the time was as wrong as it could be. Her behavior might not be treasonable according to statute, but it was unfathomably selfish, or so people thought, and they would remember, never thinking of her or the Anti-Cigarette League the same way again.

But Gaston seemed to know how much was at stake, to sense that World War I might well be a blow from which her cause could not recover. As a result, she threw caution aside, behaving with what even she might have realized was a self-defeating desperation. She looked ahead, saw that the Americans who needed cigarettes to face the strains of battle would come home from Saint Mihiel, the Marne, the Argonne Forest, and other scenes of trepidation and horror smoking more than they ever had before. Now they would have the strains of civilian life to confront; these, too, would prove taxing, and cigarettes would be no less effective in dealing with them. The soldiers would also bring back memories of the fighting, and for these the weed would also be a palliative. Gaston felt it coming, and she lashed out at the impossible odds.

Per capita consumption of cigarettes before American involvement in World War I was 134 a year. When the boys came marching home again in their various stages of shell shock and nicotine dependence, the total was 310 and climbing.

THE WAR OFFICIALLY ended with the signing of the Treaty of Versailles on June 28, 1919. Less than six months later, the Eighteenth Amendment to the Constitution of the United States went into effect. It was no longer legal to manufacture, sell, or transport alcoholic beverages anywhere in the country. The Volstead Act, named after a Minnesota congressman who was not particularly zealous in his opposition to tippling, provided for enforcement of the amendment, and a special corps of federal agents was created to crack down on scofflaws. Penalties were so severe, thought members of the temperance movement, that few people would think of testing either the law or the agents' mettle. To them, the liquor industry had been vanquished as thoroughly as had the Central Powers. The evangelist Billy Sunday was especially jubilant. "Prohibition is won," he crowed. "Now for tobacco."

But it was not to be. Within a year and a half of Sunday's boast, the legislatures of Tennessee, Iowa, and Arkansas had decided that anti-cigarette laws were an exercise in futility and repealed them. The next year, 1922, cigarettes finally overtook cigars as the most popular form of tobacco in the land. Part of this had to do with conditions in the cigar-making industry, where the work was often farmed out to immigrants, men and women who made the smokes in the same rooms in which they lived. "In one tenement," write James

MacGregor Burns and Susan Dunn, "[New York legislator Theodore Roosevelt] found two families—Bohemians, with no knowledge of English—living in abject squalor in one room; tobacco was everywhere, alongside the foul bedding, next to food." Journalists exposed conditions like these; legislators debated them; the cigar companies reeled from the effects of the publicity.

But more than anything else, it was preference that accounted for the cigarette's overtaking of the cigar, simple preference. In 1922, the former held a slight edge in popularity. The numbers would never be close again.

Still, Lucy Page Gaston did not give up. She would never give up, she vowed, although she might by now have realized that she was only going through the motions, doomed to a life on automatic pilot because her convictions were so powerfully unrealistic that they would not allow her to change course. At one point in 1920, she even began to criticize the Anti-Cigarette League, which she believed was not working diligently enough to achieve its goals. She did so publicly, loudly, and often. The league responded by firing her, and it is a rare organization that dismisses the person who brought it into existence.

Gaston was also upset at the federal government's unwillingness to support league goals. Something had to be done about this, she said, and it was she who would do it. She announced her candidacy for the Democratic nomination for the presidency of the United States. Her platform was opposition to tobacco and not a single thing else. Now that the world was at peace again, what else *was* there?

Gaston knew she had no chance, but she put up a brave front to reporters, at least at the outset. She reminded them of that resemblance of hers to honest Abe Lincoln and suggested that by virtue of physiognomy alone she should be considered a formidable challenger. Reporters laughed—that is, when they responded to her at all. One of them pointed out that the Republican candidate, Warren G. Harding, was a man of presidential bearing himself. Gaston disagreed. He has cigarette face, she declared; anybody could see that. Harding would be, in fact, the first occupant of the White House openly to indulge in the small smokes.

Gaston dropped out of the race long before election day. Most newspapers did not cover her withdrawal or the statement she released that she was now throwing her support for the Democratic nomination to William Jennings Bryan, a three-time loser in the White House sweepstakes who did not even get the nomination this time. For Gaston, the failures and frustrations were mounting.

Harding had not been in office more than a few months when Gaston decided she had to get his ear. She wrote to him, urging that he set an example for the boys and girls of America by giving up tobacco. After all, she pointed out, one of his predecessors, William McKinley, was on record as saying, "We must not let the young men of the country see their President smoking." The new president ought to take the same stand.

Harding ignored her.

Then, a short time later, a pro-leaf group in Kansas sent a carton of cigarettes to the White House, a present that the group hoped the occupant would accept as a means of showing his support for the industry. He did. "I think it is fine to save the youth of America from the tobacco habit," he said by way of explanation. "I think, however, the movement ought to be carried on in perfect good faith and should be free from any kind of hypocrisy or deceit on the part of those who are giving it their earnest attention."

Gaston had no idea what he meant. Neither did reporters covering the White House at the time. Neither does the author of this book. Sometime later, Harding did say that he would no longer smoke in public and seems to have been true to his word. But that was not enough for Gaston. She wanted him to be a positive and active role model for the anti-tobacco cause. Instead, the president turned his attention elsewhere—to immigration, taxes, trade, the League of Nations—never again to speak ex officio on the subject of tobacco and probably not giving it much thought in private, either.

Harding died before he could seek reelection. His vice president, Calvin Coolidge, finished the term and in 1924 ran for the nation's highest office on his own, defeating the Democrat John W. Davis in a landslide. That same year, Americans would smoke 73 billion cigarettes, which worked out to 600 for every man, woman, and child, and was eighteen times as many small smokes as they had consumed a quarter of a century earlier, when Gaston founded the Chicago Anti-Cigarette League, so confident she could make a difference.

Her last years were painful ones. "With no regular salary," Cassandra Tate tells us, "Gaston was forced to rely on handouts from relatives and charities. Her brother Edward said later that she often walked for lack of money to take a streetcar." She lived mostly on graham crackers, which were then thought to be a health food, and drank a daily five-cent glass of milk at a lunchroom. On holidays, the Salvation Army, not holding a grudge, sent her a basket of fruits and vegetables and meats.

But she continued to denounce the weed. "We are out to put the cigarette business out of business," she said for the record, and she "harangued women smokers; collared boys she saw smoking on the street corners; and handed out news releases and gentian root ... to reporters." She never let up, never gave in to a reality that she believed was both dangerous and demeaning.

AND THEN CAME AUGUST 20, 1924, a particularly hot and humid day in Chicago, with a damp breeze blowing in off Lake Michigan. It was a day on which Coolidge was campaigning for the presidency and Leon Trotsky was predicting worldwide civil war and Babe Ruth hit his thirty-ninth home run of a season in which he would hit forty-six. It was also the last day of Lucy Page Gaston's life. Six months earlier, she had been struck by one of the streetcars she could not afford to ride after attending an anti-cigarette rally. She was pinned beneath the car for almost an hour, with most of the weight landing on her neck. The trauma was thought to have contributed, at least to some extent, to the terrible irony of her death, at age sixty-four, from cancer of the throat.

Her funeral, a few days later, was sparsely attended, although those who did show up spoke of the deceased in most laudatory terms. She was dedicated, conscientious, selfless. She was one of a kind and would sorely be missed. A surprising number of newspapers were complimentary, as well, including journals in the Midwest and on the Pacific coast, in Mobile, Alabama, and Boise, Idaho, and Ann Arbor, Michigan. Even a trade publication called the *Tobacco Leaf* acknowledged Gaston's "fine character and splendid ability." The *San Francisco Call* conceded her stubbornness and unpopularity but then asked a plaintive question: "Haven't you a little admiration to spare for Lucy Page Gaston?"

A surprising number of people, it seems, did.

NINE

The Last Good Time

BY 1927, all fourteen states that had either banned or limited cigarette sales between 1895 and 1921 had changed their minds and given the smoke at least partial freedom again. No new laws were added by other states. In fact, when an anti-smoking bill was being considered in Utah, a member of the legislature got so angry about it that he threatened to introduce a proposal of his own to forbid the public sale of corned beef and cabbage. He said the fumes from that stuff were as distasteful to some of his constituents as those of the weed were to others. Neither, he implied, was more than a temporary annoyance; neither required legal redress.

Local regulations remained in effect in a variety of places, though, especially those forbidding the purchase of cigarettes by minors, but enforcement became ever more haphazard as the years went on. Soon the laws had no more credibility, and no more muscle behind them, than the law intending to eliminate booze. In fact, the extent to which Prohibition was being flouted made tobacco reform seem more of a fantasy than ever. You cannot legislate appetites, people were beginning to say, and the evidence seemed persuasive.

Looking back, it all has the feel of inevitability to it. The times were dynamic, electric; the cigarette was made for the peace that followed World War I even more than it had been made for the fighting. There was Ruth hitting home runs and Dempsey hitting Carpentier and Robert Goddard launching a rocket a quarter of a mile or more into the air. There were flagpole sitters launching themselves, perching ten or twenty or thirty feet off the ground on tiny platforms in the shadows of skyscrapers, of which there was an ever-increasing number, especially in the flagpole sitting capital of New York. There was Lindbergh crossing the ocean and Gertrude Ederle crossing the

English Channel and people like Hemingway and Fitzgerald and Lewis and Stein and O'Neill and Picasso crossing virtually every barrier of art—and, some said, decency—that had ever existed. Motion-picture shows were a *new* art, and the very idea that there could be such a thing after all these centuries of poems and plays and novels and songs and paintings and sculptures was miraculous in itself.

There were Sacco and Vanzetti claiming innocence and the perpetrators of Teapot Dome claiming innocence and Leopold and Loeb echoing Nietzsche and Darrow echoing Darwin and the Ku Klux Klan coming back to life and a grand jury indicting the blackened White Sox, who had sinned against the ideals not just of sport but, because the sport was baseball, of all America. Had there ever been such stories as these before—such controversy, such drama and furor, such flesh and blood in journalism? Had there ever been headlines so thick and black, newsboys' cries so shrill? Stories like these got into a reader's bloodstream, coursed through him, affected him, *in*fected him.

Automobile sales tripled from 1917 to 1923. Radio sales increased *500,000 percent* in just four years, from 1920 to 1924. Between 1922 and 1928, "the seven fat years," the gross national product increased 40 percent, and per capita income shot up almost 30 percent.

In 1927, President Hoover's image was beamed from Washington to New York by something called television. Hemlines climbed up the thigh, and the women who raised them were also bobbing their hair and chucking their corsets and dancing the Black Bottom and the Charleston and taking Freud and those fascinating new theories of his, oh, so seriously as they held on to their men for dear life as the men held on to their stock tickers for dear life, watching the tape unspool and grinning wildly, riding the crazy ascent of the bull market into the stratosphere, "into regions once considered as remote as the moon," most of the riders certain that the market was the single exception to the rule, the lone amendment to the constitution of physics, the one thing that went up and would never, ever come down.

And the cigarette—the easily replaceable, instantly rechargeable, immediately gratifying cigarette—accompanied the era as perfectly as a mad drummer or a howling saxophone. Popular fiction reflected the times. Now it was the heroes who were smoking coffin nails and the villains who lit the cigars. Pipes were antiques, plug a means of stamping a character as a rube. Such a turnabout it had been for the small smoke in the United States, where the fine brown leaf had been one of the founding graces. The cigar was a relic; the age of immediacy had arrived. The tracks were being laid for the rat race—long live the cigarette!

BUT THE COMPANIES who made it wanted more: more customers, more sales, more respect. They did not sit back and accept the zeitgeist's bounty. They worked hard, built on their successes, succeeded more, built on those—a rat race of their own.

They went after the lazy smoker, one firm bringing out a cigarette with a built-in striker at the end so that a person would not have to reach all the way down into his pocket for a match. They went after the literate smoker, another firm providing a microscopically tiny volume of Shakespeare in every pack. They went after the sports fan, some companies offering pictures of athletes with autographs for whose authenticity no one could vouch. They went after the flesh worshiper, a few brands including pictures of singers or actresses or bosom beauties of some other sort who posed seductively, smiled teasingly. They went after the happily married male tobacco clientele, another brand enclosing small pieces of silk in their packs, which the men were instructed to give to their wives "to piece out pillows for their horsehair sofas." They went after the college crowd, yet another brand handing out a Yale pennant with the purchase of each pack. And they went after children, boys and girls, teenagers and even younger, with Sweet Caporals, one of the most popular smokes of the day, adopting the slogan, "Ask Dad, He Knows."

It was a time that called for superlatives. Americans were consuming more cigarettes than ever before. As a result, more companies were manufacturing cigarettes than ever before. As a result of *that*, the potential for profits was greater than ever before. It was crucial, then, for one maker to distinguish itself from another, to make its product seem somehow unique and therefore create a demand for it that could not easily be satisfied by or transferred to another brand. And so advertising became more important than ever before. "Modern business," said President Coolidge, perhaps modern business's leading spokesman of the period, "constantly requires publicity. It is not enough that goods are made—a demand for them must also be made."

As it happened, the palmy days of the cigarette trade coincided with, and perhaps even helped to bring about, a change in the very philosophy of merchandise hawking. In his influential 1923 book *Scientific Advertising*, Claude Hopkins explained, "The product itself should be its own best salesman. Not the product alone, but the product plus a mental impression, and atmosphere, which you place around it." In other words, one no longer sold an item simply by telling people how well it worked or how reasonably it was priced. Now one sold the product in large part by the associations created on its behalf.

The Reynolds Tobacco Company already knew that. In fact, Hopkins might well have been referring to a specific Reynolds ad, and others like it, when he wrote the preceding. On December 12, 1914, the company purchased two full pages in perhaps the nation's most prestigious magazine, the *Saturday Evening Post*, to introduce its new smoke. "The Camels Are Coming!" is what the copy said and went on to point out that the cigarette featured a uniquely mild blend of tobaccos. But the ad got no more specific than that; the rest of it was mental impressions and atmosphere, a carefully chosen color scheme and just the right size and design of typeface. It was a totally new approach for a brand new brand, and both Madison Avenue and the tobacco industry knew the risk.

Camels, though, were a hit. Sales started out briskly and improved every week for the first several months, while the new brands of other companies were faltering, a few of them being discontinued virtually within days of their first being marketed. Assuming that advertising made the difference, Reynolds was quick to put its money where its faith was, in 1916 spending the unprecedented sum of 2 million dollars for appeals to consumers in various newspapers and magazines. As a result, at least in part, the following year Camels accounted for 35 percent of all cigarettes sold in America. In 1920, the first full year of peace after World War I, and with the majority of America's tobacco-crazed fighting men back home battling their foxhole demons as they fretted about the future, the Reynolds ad budget tripled. Three years after that, the Camels' market share was a remarkable 45 percent.

But just as advertising helped to legitimize cigarettes, so did cigarettes return the favor. The tobacco companies were a building block of the modern ad industry, one of the businesses on which its fortunes were founded. Even before the industry *was* modern, in fact, James Duke was getting it off to a smashing start by spending 20 percent of his revenues on advertising, an unprecedented amount. And it was probably on the weed's behalf that advertising made its most dramatic change ever. It developed something called the Unique Selling Proposition, or USP, which, in a less pretentious term, was simply the use of a slogan, such as "The Camels Are Coming!" instead of a densely worded paragraph. From now on, an ad would no longer consist of one or two small pictures and a hundred lines of copy in a daily paper or magazine, the whole thing looking like an essay on the minutiae of Versailles reparations. Instead, it would be big pictures and a couple of phrases and a lot of space between them, with the copy simple enough to be catchy, catchy enough to be repeated, repeated enough to brand itself onto the brains of smokers and buyers of cigarettes from one end of the country to the other.

There have been a lot of memorable USPs in the cigarette industry over the years. Perhaps the second of them kept sales shooting up for the new Reynolds brand. "I'd Walk a Mile for a Camel!" the ads boasted, and many Americans, it seemed, actually would.

But it is probably true that no single ad or advertising campaign gave as much of a boost to the tobacco companies at this time as an act that was probably spontaneous, that the companies knew nothing about in advance, and that cost them not so much as a single cent.

Shortly after returning to the United States after his epic flight to Paris, Charles Lindbergh made a public show of lighting a cigarette and puffing on it in exaggerated contentment. He inhaled, he smiled, he exhaled, he kept inhaling and exhaling and smiling. Reporters took note; photographers took pictures. It was an uncharacteristic display for the normally reticent aviator, but he thought he had been provoked, and there were few things in life that bothered him more.

The provocation was a statement by a women's group, a few weeks earlier, that urged children not to smoke because "Lucky Lindy" did not. The implication was that boys and girls would improve their own chances for heroic behavior by refraining from the leaf. Simple as that, innocent as that.

But the women's group had not asked Lindbergh's permission to use his name, and he, one of the touchier heroes America has ever known, was much displeased by their negligence. He was also displeased by the "Lucky Lindy" nickname, which always made his molars grind, not to mention the countless other displays of public adulation that had been aimed at him since his landing in Paris. So the aviator decided to thumb his nose at the do-gooders. "I won't be played for a tin saint," he declared, overreacting greatly, and in that instant, puffing away for the press, he became the greatest celebrity endorser of tobacco up to that time, and possibly ever.

BUT IT WAS NOT READERS of Shakespeare that the cigarette companies most wanted to reach in the 1920s. It was not Ivy Leaguers or sports fans or children too young to know better. It was not oglers of the feminine form. Rather, it was possessors of the feminine form whom the industry coveted, women themselves, the same gender that was throwing away its corsets and taking up its hemlines and drinking bootleg hooch out of hip flasks and, all in all, behaving in ways that their grandmothers would never have recognized, much less found satisfactory. It was the same gender that had expanded its horizons by aiding the war effort as nurses and volunteer workers overseas; the same gender that had taken the places of men in some of the factories and offices at home and was, as a consequence, not willing to return to more restrictive roles with the peace. It was women who were the next frontier for the American tobacco industry, the next potential eruption in profits.

But the accent was on the "potential." Cigarettes had undergone such a drastic reversal in World War I, had been so completely masculinized by it, that those who manufactured them feared they had lost the woman forever, that she had become inadvertently alienated despite her natural longing for a smaller, gentler, quicker smoke. She might have admired the efforts of the doughboy and the fighter pilot, might have gratefully accepted their sacrifices, but did she want to light up as they did?

Yes, answered cigarette makers on her behalf, but at this stage the response might have been more hopeful than confident.

According to J. C. Furnas, "The cigarette smoking woman in America is traceable back to fast theatrical circles of the mid-1800s." He continues:

> But her entrance under sponsorship from elegant society may have waited until the 1880s, when Mrs. Burton Harrison, well placed among the best people, mentioned in a novel, *The Anglomaniacs*, how after-dinner cigarettes had "swept like a prairie fire over certain circles" of ladies after being "introduced by a Russian lady of rank in Washington."

Mrs. Harrison probably exaggerates. Cigarettes might have gained a certain acceptance in the upper reaches of society, but they did not sweep like a prairie fire over that or any other segment of the female population, not back then. Nor did the male population assume such a thing. Relations between the sexes were awkward on this point, with women not sure how to behave in a social setting that featured men smoking, and the men uncertain what to expect of the women.

Some of the tension showed one evening in a fashionable restaurant in an unidentified American city, where an Englishwoman, as the story goes, was dining and hoping to be noticed. The headwaiter showed her to her table, held the chair for her, and watched for a moment as she regally slid into place and surveyed the room. Then she asked him, in the manner of one who already knows the answer to a question and simply wants confirmation, whether ladies were permitted to smoke in his establishment. She might already have been reaching for her purse and the pack of cigarettes within. "Ladies may, madam," came the reply, the tone itself a bit regal, "but ladies never do."

It is for this reason that women did not light up—or, for that matter, consume an alcoholic beverage or utter an expletive—in the *Saturday Evening Post*'s short stories of the time.

In 1919, awkwardness became institutionalized at the Scribner's publishing house when women were hired to operate a new batch of typewriting machines. The male employees of the firm were told not to smoke in the vicinity, the assumption being that the women would not approve. And, in fact, some did not. Others, though, were fond of the weed and therefore opposed to the policy, while still others thought that men, in acceding to it, were being condescending, and that in truth they pitied the typewriter operators their exclusion from so basic a mortal pleasure as tobacco.

In some cases, that exclusion was a matter of law. A decade and a half earlier, a New York judge had sentenced a woman to thirty days in jail for allowing her children to see her smoke. A Secaucus, New Jersey, schoolteacher lost her job "for impairing the morals of her pupils by smoking in public." A Lockland, Ohio, man won a divorce from his wife on the grounds that she was a "cigarete [*sic*] fiend." A mounted policeman in New York, finding "a lady cowering in the rear of an automobile, trying to sneak a cigarette," said, "You can't do that on Fifth Avenue!" And a Democratic congressman from Mississippi sponsored a bill in 1921 to forbid ladies from smoking in the company of gentlemen in Washington, D.C. "The women can not save this country by trying to get down on a level with men," he explained. "They must pull the men up to where they are, and they can not do it by smoking cigarettes."

But it was now the Roaring '20s, and more than anyone else it was the newly emancipated American woman who made them roar. They did not care about pulling men up to their level. They did not care about men who thought smoking was unladylike. Or perhaps it is more accurate to say that being ladylike, in the traditional sense, was something they did not care about anymore.

A Chesterfield ad from early in the decade illustrates the point: another USP, another set of associations. It shows a man and a woman seated by a lake with a full moon overhead. The man lights a cigarette. His companion looks at him with longing in her eyes. The copy reads: "Blow Some My Way."

It was an eye-catching tableau. But by the time it appeared in the mass-circulation magazines, it might already have been out of date. Women wanted to do more than merely sniff the smoke that a man exhaled; they wanted to produce their own quantities of it, large quantities, and blow it *his* way. In 1921, the Nineteenth Amendment to the Constitution gave them the right to vote. To many American women, the amendment had a symbolic as well as a literal significance; it meant that they were now to be regarded as the equal of men in *all* ways, not just at the ballot box.

As a result, women sought more freedom in the workplace, fewer limitations at home, more social and cultural opportunities than had been available to them before. And they went further. They took over the smoking room at the Lantenengo Country Club in John O'Hara's *Appointment in Samarra*, "muscling in on" it. And, according to Frederick Lewis Allen, they "now strewed the dinner table with their ashes, snatched a puff between acts, invaded the masculine sanctity of the club car, and forced department stores to place ornamental ash-trays between the chairs in their women's shoe departments."

There was a huge increase in the number of female cigarette smokers—or, at least, in the number of them smoking publicly, even brazenly. In studying the figures, the psychoanalyst A. A. Brill came to this conclusion: "More women now do the same work as men do. Many women bear no children; those who do bear have fewer children. Feminine traits are masked. Cigarettes, which are equated with men, become torches of freedom." Then, perhaps catching himself, realizing that he sounded insufficiently Freudian for the era, he shifted gears. "Smoking," he said, "is a sublimation of oral eroticism; holding a cigarette in the mouth excites the oral zone."

Cassandra Tate, in her book *Cigarette Wars*, agrees:

> Particularly when smoked by women, cigarettes seemed to unleash a disquieting sexuality. Although there is an element of sensuousness in the use of any kind of tobacco (the mouth and hands being intimately involved whether it is chewed, snuffed, or smoked in pipes, cigars, or cigarettes), the effect seems more pronounced with cigarettes. Perhaps this has something to do with the frequency with which cigarettes are brought to the mouth, with the smoke being deeply inhaled, suggesting a titillating degree of intimacy.

That lost lady of Willa Cather's, her husband so polite about cigar smoking in mixed society, could hardly believe it. Women exciting their oral zones? Women demonstrating a disquieting sexuality? This was not the world she knew, or the one in which she wanted to reside. She expressed her bewilderment to a young male admirer:

> "And tell me, Niel, do women really smoke after dinner now with the men, nice women? I shouldn't like it. It's all very well for actresses, but women can't be attractive if they do everything men do."
>
> "I think just now it's the fashion for women to make themselves comfortable before anything else."
>
> Mrs. Forrester glanced at him as if he had said something shocking.

She would have been even more shocked if she had been told of events at Smith College, a more-than-proper school for girls, half a century old, in Northampton, Massachusetts. Smith had long prohibited smoking on campus, finding it a habit unworthy of the daughters of the kinds of people who could afford to send their offspring to such an exclusive place. But early in the '20s, the student government bent before the pressures of the age and overturned the ban. The vote was almost unanimous.

It fell to the school's president, William Allan Neilson, to announce the change of policy to the student body at an assembly. But would he? There were fears that Neilson might veto the decision. Although some of the girls thought highly of him, to others he seemed the kind of dour, ascetic fellow who would not only be opposed to smoking himself but would delight in denying the pleasure to those he supervised. As he strode to the stage and planted himself behind a podium, with the members of the student government seated behind him, the fears seemed validated. Smoking, he said, leaning into the microphone, his voice deep and halting, "is a dirty, expensive, and unhygienic habit."

The girls behind him gasped.

"To which I am devoted," Neilson said, and the whole auditorium roared its approval, many of those assembled reaching for their packs and matches, no longer needing to hide them, lighting up, and drowning one another in hot gray clouds of freedom.

Just their luck that in short order two fires broke out in Smith dormitories, and the careless handling of cigarettes was suspected in each. Restrictions went back into place; personal liberties gave way to campus safety.

In 1923, an estimated 6 percent of all cigarette smokers in the United States were women. New ad slogans were being written for them: "Women—when they smoke at all—quickly develop discerning taste. That is why Marlboros now ride in so many limousines, attend so many bridge parties, and repose in so many handbags." And new kinds of smokes were being invented for women, such as "cocarettes," which were "a soothing blend of refreshing Colombian coca-leaf and the lightest Virginian tobacco, specially blended for the Lady's need." As a result, by 1929, the percentage of female smokers had doubled, and it was estimated that women consumed a total of 12 percent of all cigarettes sold in America. Two years later, they were consuming 14 percent.

But the tobacco companies were far from satisfied. They wanted more: not scattered victories but total conquest; not a campus here, a ladies' club there, a few unaffiliated members of the fairer sex somewhere else, but an entire nation, the unconditional surrender of all opposition to distaff smoking from one end of the United States to the other.

In fact, one tobacco executive said that he thought women should smoke cigarettes as often as they ate candy. It was an ambitious goal for the time, and even he did not know whether it was possible. The advertising campaign that tried to make it so was not only one of the most controversial in the history of Madison Avenue but a cultural landmark on all American avenues, for all American smokers.

THE THIRD MAN to head the American Tobacco Company took over in 1925, succeeding his father, who had in turn succeeded the founder, old James Duke. The new guy's name was George Washington Hill, but he bore little resemblance to the first president of the United States, in either demeanor or priorities, integrity or thoughtfulness. Or even height, for that matter. Washington stood six-feet-two in his stockings; Hill was a "rawboned, diminutive figure," coarse and hearty, self-absorbed to an almost self-mocking degree, the kind of man who makes entertaining reading to later generations but does little to endear himself to those who are cursed with his daily presence, in either the office or the home. You could tell just by looking at him. There was something about the jut of his jaw—too pronounced; about the intensity of his eyes—too piercing; about the sweep of his eyebrows—as bushy as undergrowth after long and musty rains.

George Washington Hill ruled the American Tobacco Company like a plantation owner ruling his acreage, sitting "behind a desk that seemed to take up half of the corner office he had inherited from his father. Wearing a tilted Stetson with fishhooks protruding from the brim, he'd hold court in a plainspoken reasonable manner. Then, without warning, he would explode in a tirade at his stenographer, advertising executive, or anyone else who happened to be there." He liked to be driven around New York in a Cadillac convertible: chauffeur at the wheel, bodyguard beside him, both attesting mutely to the passenger's importance. And he expected further attesting from the people with whom he came into daily contact; he wanted reverent tones, subservient demeanor, constant agreement.

As far as anyone knows, when Hill was a boy he did not dream of one day being a fireman or a soldier or an athlete. He wanted to be a tobacco executive, and started learning all that he could about the leaf as a youngster. He learned the economics of smoking, the politics of it, the horticulture of it, the chemistry of it, the marketing of it, even the damage control of it. He wanted to grow up and work for his dad, perhaps even take over for him one day. Hill lived, breathed, and ate the large brown leaf, almost literally. He also raised it, which seems to have been as close as he ever came to a hobby, tending to some tobacco plants he owned, not for commercial gain but for his own enjoyment, in a small garden at home.

In later years, also at home, Hill had "radios in every room [that] bathed him in the sound of his own cigarette commercials." It was he who created the signature American Tobacco brand, Lucky Strike, promoting it from its lowly status as a pipe blend that no one was buying anymore and then pitching it tirelessly, endlessly, tediously. One of its first Unique Selling Propositions was "Lucky Strike Means Fine Tobacco," and you needed only to live in the United States at the time, not necessarily to be a smoker, to have heard the phrase as often as you heard your children's names. After a while, it became abbreviated in the ad copy: LSMFT, just the initials, and everyone knew what they stood for. Hill even named his dachshunds after the cigarette: One was Mr. Lucky and the other Mrs. Strike.

More than just a successful brand, though, Lucky Strike was the salvation of American Tobacco. In 1911, the company had been found guilty of monopolistic practices, having created, under Duke, the so-called Tobacco Trust. According to the U.S. Supreme Court, which ruled in 1911, the company had sought "dominion and control of the tobacco trade, not by the exertion of the ordinary right to contract ... but by methods devised in order to monopolize ... [and] by driving competitors out of business." As a result, the court declared, American Tobacco was broken up into three separate entities, thereby loosening its stranglehold on the industry, and James Duke decided to retire.

At which point George Washington Hill's father took over. Fourteen years later, it was the son's turn, but not until he came up with Lucky Strike did the company truly begin to prosper, in no small measure because of money spent on advertising. It was said that, at the time of his death in 1946, Hill had spent more than a quarter of a billion dollars promoting Lucky Strike cigarettes, the greatest amount ever allocated up to that time trying to sell a product of any kind.

It was Hill who thought that cigarettes ought to be as ubiquitously enjoyed as candy, and he explained once how the idea came to him. He was riding home from his office one day, he said, and, caught in traffic, found himself getting angrier and angrier at the delay, then getting angrier and angrier at himself for his reaction. Determined to cool off, to get something constructive out of the delay, he began to check out his surroundings, idly at first, then with more and more attentiveness. He said that he

> looked at the corner, and there was a great big stout Negro lady chewing on gum. And, there was a taxicab—it was in the summertime—coming the other way. I thought, I was human and I looked, and there was a young lady sitting in he taxicab with a long cigarette holder in her mouth, and her skirts were pretty high, and she had a very good figure. I didn't know what she was smoking; maybe she was smoking a Camel.

> But, right then it hit me; there was the colored lady that was stout and chewing, and there was the young girl that was slim and smoking a cigarette—"Reach for a Lucky Instead of a Sweet."

One wonders what Hill would have thought had the black woman been svelte and puffing away and the white one obese and chomping on her gum.

Regardless, the expression "Reach for a Lucky Instead of a Sweet" went straight from Hill's lips to the USP Hall of Fame. Lucky Strike no longer meant fine tobacco; it was not a bold enough claim. Now it meant a course at mealtime, an actual part of a well-balanced diet. "Reach for a Lucky Instead of a Sweet" was the "Does She or Doesn't She?" or "Just Do It" of its day, not just a motto but a mantra in scores of ads in print and on the radio and, even more insidiously, an addendum to the vernacular, the phrase incorporated into the conversations of smokers and non-smokers alike, who uttered it, or a knowing variant, to demonstrate a certain pop-culture savviness.

The credit for this, if credit is the proper term, belongs only initially to Hill. For the most part, it was the doing of another man, someone who came along later and put flesh onto the slogan's bones, making it the basis of a corporate strategy that was as wide-ranging as it was perversely brilliant, as influential as it was bereft of logic. According to the title of a biography by Larry Tye, Edward L. Bernays was *The Father of Spin*, one of the inventors of the craft of public relations and certainly as creative a practitioner as the field has ever known. The title of Bernays's own book, *Crystallizing Public Opinion*, published in 1923, summarizes his mission statement.

In this case, the mission, implausible if not actually out of the question, was to turn cigarettes into a substitute not only for gum but also for chocolate bars and saltwater taffy, licorice and caramel, ice cream and pastries. Where once American women had wanted something cool and sweet at the end of a meal, they were now to crave a hot, bitter combustible. That was what Hill said to Bernays. What Bernays said to Hill in response was something along the lines of: Can't imagine anything more sensible than that.

The first thing he did, once Hill had agreed to Bernays's outsized fee, was to approach a photographer friend and tell him "to ask other photographers and artists to sing the praises of thin." This they did, hitting one crescendo after another as they pointed out not only that thin was beautiful, but that cigarettes were a terrific way to achieve a narrow waist and flat stomach. No more of this rotund little cutesy-pie Gibson Girl look; it was time now for lean and angular.

Before long, the theme was being repeated by society columnists, feature writers, and the average American woman, who, having read what the others were writing, assumed that they were speaking the gospel truth about the culture.

Then Bernays, as always operating slyly, not mentioning Lucky Strike by name or even admitting to people that a tobacco company was his client, began to work

> directly to change the way people ate. Hotels were urged to add cigarettes to their dessert lists, while the Bernays office widely distributed a series of menus, prepared by an editor of *House and Garden*, designed to "save you from the dangers of overeating." For lunch and dinner they suggested a sensible mix of vegetables, meats, and carbohydrates, followed by the advice to "reach for a cigarette instead of a sweet."

Since Lucky Strike was the only brand offering itself as an improvement on sweets, Bernays could make his point without actually stating it. Sly, he had long since discovered, was much more effective than overt; it set off fewer alarms and therefore sold more merchandise.

The deviousness of his ingenuity knew no bounds. Bernays oversaw an advertising campaign that featured, among other things, an ad referred to in the trade as "The Grim Sceptre," in which a woman with a double chin was implored by a line of copy to light up a Lucky Strike rather than break open that next box of candy. Less directly, Bernays urged that kitchens be constructed in such a way as to provide space for cigarettes on their counters and shelves. He urged, in other words, a society which made cigarettes so handy that it took more of an effort to avoid them than to pick one up and strike a match.

The campaign became a legend. Insiders spoke of it in terms of awe and envy, and several college students and recent graduates of the time were quoted in newspaper reports as saying they had been so inspired by it that they changed their previous career plans and converted themselves to flackery. Bernays's work was written about extensively in both trade journals and textbooks and studied as if it were a chemical formula that unlocked new possibilities in the physical world.

It did unlock new possibilities for American Tobacco. In 1928, the first full year of the "Reach for a Lucky Instead of a Sweet" campaign, the brand's sales showed a greater increase than the sales for all other cigarettes brands in the United States combined. Lucky was, in fact, second only to Camel in popularity among cigarette smokers. George Washington Hill was delighted. He got just a little bit fuller of himself. And, two years later, in 1930, when Lucky overtook Camel and the former president's daughter Alice Roosevelt Longworth expressed her satisfaction with the product by doing an ad for Lucky, Hill practically popped the buttons of his vest all over the back seat of his limo.

But the impact of the campaign may also be measured in terms of the responses it drew from the various American companies that manufactured confections. They were sharp, angry, sarcastic. They were laced with invective, mounted in boldest type. They were not, however, prompt. The makers of cakes and candies and cookies had not been prepared for the Hill–Bernays assault, and understandably so. It was one thing for a company to criticize another company that made the same product—that was competing for the same share of market. But this was one entire business criticizing a different *set* of businesses, apples claiming to be superior to oranges, an unheard-of thing in the world of advertising.

What the sweets companies lacked in speed of response, however, they made up for in quantity. They paid for billboards, posters, flyers, ads in all available media—a virtual cacophony of retaliation. A chain of candy stores in New York included the following warning in its ads:

> Do not let anyone tell you that a cigarette can take the place of candy. The cigarette will inflame your tonsils, poison with nicotine every organ of your body, and dry up your blood—nails in your coffin.

Lucky Strike countered with an endorsement from the Broadway star George M. Cohan. "Lucky is a marvelous pal," he said. "The toasted flavor overcomes a craving for foods which add weight.

Whereupon the candy industry called a meeting of executives "representing some two dozen manufacturers and trade associations … to discuss ways and means of quashing the 'fattening sweet' menace."

Following which the singer Al Jolson announced: "I Light a Lucky and Go Light on the Sweets."

Which, in turn, inspired the dessert and snack makers of America to inform the populace that: "Sweets 'fixed' saliva … which not only deadened appetite, but lessened the urge for cigarettes."

To which Lucky Strike replied: "A reasonable proportion of sugar in the diet is recommended, but the authorities are overwhelming that too many fattening sweets are harmful and that too many such are eaten by the American people."

Thereby bringing the National Confectioners' Association into the fray with: "Don't neglect your candy ration!"

And so it went, back and forth like that, the pot and the kettle blasting each other on charges of blackness for month after month, with constantly increasing dissonance, higher financial stakes, and no regard for truth or accuracy or even common sense.

Finally, a third party, the Lorillard tobacco company, figured out a way to capitalize on the dueling claims between smoke and sweet and, by so doing, offered a nicely ironic, if totally unintended, commentary on their preposterousness. Lorillard took out an ad of its own and ran it countless times in countless publications: "Eat a Chocolate, Light an Old Gold. And Enjoy Both! Two Fine and Healthful Treats!"

But Lucky Strike was the real winner in the great tobacco–candy showdown of the 1920s and early '30s. It is unclear whether masses of people reached for a Lucky instead of a sweet, but it is certain that they reached for a Lucky more than they reached for any of its competitors, Camel included.

Still, as late as 1934, George Washington Hill was sitting behind his big old fortress of a desk, grumbling. Even now he had not attracted enough female smokers to content himself. Even now he was seeing too many women on the street with their lips vacant, too many women at social gatherings without a stoked-up Lucky in their grasp. Something was wrong, and, unable to figure it out on his own, Hill commissioned a poll. The results could not have surprised him more. It seemed that American women had nothing against the cigarettes per se; they tasted fine, were easy on the intake, were priced competitively. It was the pack, of all things—the Lucky Strike pack with the red bull's-eye and the green trim that offended them. The green, in particular, was a problem. It clashed with their clothes, or so a good number of women told pollsters and, not only that, did not please their eyes. They had no desire to reach into a package that dissatisfied them aesthetically for a smoke that satisfied their taste buds. The solution, the pollsters in turn told Hill, was obvious: Come up with a different color.

Hill's response, at least to those who knew his voluble nature, was equally obvious: "I've spent millions of dollars advertising the package. Now you ask me to change it. That's lousy advice." And he simply would not do it.

Once again Hill summoned Bernays, and once again it was Edward L. to the rescue. He agreed that a change had to be made but suggested something far more radical than remaking the Lucky Strike pack, which might have struck him as insufficiently challenging. Bernays told Hill he would, in effect, remake the fashion outlook of American women. He would see to it that they rethought their wardrobes, their notions of style, their very concepts of color coordination and visual enrichment, if not even their place in the cosmos and the role of a Creator in establishing the ultimate purposes of mortal existence. He would make them, in short, keen on green.

George Washington Hill told his man to start—or rather, keep on—manipulating.

Bernays began by organizing a society event to end all society events, a splendiferous occasion called the Green Ball, at which the most high-toned, trend-setting, envy-rousing women on the entire East Coast of the United States would not only wear dresses of the title color but would accessorize them with "green gloves and green shoes, green handkerchiefs, green bandeaux, and yes, green jewelry."

Then, with Hill cackling in the background, rubbing his hands together and imagining blacker and blacker ink in the company ledgers, longer and longer limousines for his drives around town, Bernays and an accomplice

> invited fashion editors to the Waldorf for a Green Fashions Fall Luncheon, with, of course, green menus featuring green beans, asparagus-tip salad, broiled French lamb chops with haricots verts and olivette potatoes, pistachio mousse glace, green mints, and crème de menthe. The head of the Hunter College Art Department gave a talk entitled "Green in the Work of Great Artists," and a noted psychologist enlightened guests on the psychological implications of the color green. The press took note, with the *New York Sun* headline reading, "It Looks Like a Green Winter." The *Post* predicted a "Green Autumn," and one of the wire services wrote about "fall fashions stalking the forests for their color note, picking green as the modish fall wear."

Under Bernays's direction, women were further advised that green would be an appropriate color for summer, as bathing suits of that hue would go quite nicely with their suntans, "from the first strawberry flush to the last Indian brown."

Edward L. Bernays was playing American women as if they were musical instruments and he a virtuoso, plucking their strings with such charm and cunning and subtlety that they were not only unaware of his touch but probably would have been bemused by it if they had ever found out, even flattered by the attention, the lengths to which he had gone to make his case. A man does not trouble himself like that for a woman unless she matters to him very much, and the reason that the woman in green mattered as she did to Bernays is that he knew in his heart she was ready to be a cigarette smoker, *eager* to be a cigarette smoker. All she needed was for a trustworthy gentleman to come along and tell her precisely which brand to favor with her cash and loyalty. Bernays trusted himself implicitly; it seemed only reasonable to him that others would, too.

The father of spin lived a long time—more than a hundred years—and his deceits, so cleverly conceived and laboriously rationalized and smoothly executed, provided all manner of services for all manner of clients, from shaping the public perceptions of presidents Calvin Coolidge and Herbert Hoover to commemorating the fiftieth anniversary of the electric light to helping overthrow the socialist government of Guatemala on behalf of the United Fruit Company. Bernays was an equal-opportunity horn tooter. He expressed no regrets about either his specific actions or the field that, under his leadership, became one of the defining vocations of the image-obsessed twentieth century. In his view, the crystallizing of public opinion was not so much an occupation as a service, a boon that he graciously provided to the common weal. He surely agreed with the public-relations historian Scott M. Cutlip, who said, "Bernays emphasized ... that the public relations man's ability to influence public opinion placed upon him an ethical duty above that of his clients to the larger society."

Actually, Bernays did have one regret. But it would be many years before he felt it, and as he thought it over, working the matter through the labyrinthine corridors of that mind of his, he would let himself off the hook far too easily.

SOMETHING HAPPENED to America late in the 1920s and throughout the '30s that no one who lived through it would ever forget. In fact, so formative an event, or set of events, was it that even the children of the victims would bear scars, and the children's children would hear tales that they could scarcely believe were true in a land of plenty like the United States. It did not just occur on Black Tuesday, as people sometimes think of it now. The stock-market crash was so complete, so cataclysmic, so far-reaching in its consequences that the very act of plummeting took several days to happen, starting on Black Thursday, October 24, 1929, and not reaching its nadir until Tuesday, October 29, with the unprecedented sell-off of 16 million shares of stock and the similar forfeiting of at least as many dreams.

And that was just the crash itself. The effects would last much longer, year after year after year, until the United States declared war on Japan and Germany in 1941 and the American economy was forced to revive itself to meet the challenge of military engagement. It was not just "a little distress selling on the Stock Exchange," as the supercilious financier Thomas Lamont characterized the events of late October 1929 on Wall Street; it was a great deal of distress, more distress than most Americans had ever experienced before and would ever feel again, and it became known as the Great Depression.

Between 1930 and 1933, the value of industrial stocks in the United States fell almost 8 percent. Banks failed; companies went bankrupt; the flow of goods and services was disrupted and in some cases severed altogether. Some people found that their savings had evaporated, every penny that they had ever put away. Many lost their jobs or hung on to them at greatly reduced salaries, able to do nothing except go broke a little more slowly than those who were out of work.

As a result, many Americans ended up living in ramshackle circumstances, some losing their houses and apartments and unable to find new quarters. "Emaciated children who never tasted milk wandered the streets," writes C. Vann Woodward, "some shoeless in winter, too poorly clad to go to school. Milch cows dried up for lack of feed, and starving horses dropped in their harnesses."

Some human beings dropped, too, although a more common fate for the era's victims was to develop the kinds of illnesses and handicaps that are the consequences of want and malnutrition. It is said that in 1932 alone, at least 25,000 families, no longer having a place to call their own, were drifting from one town to another, begging for jobs or, failing that, for handouts of food, clothing, or shelter. Their sons—some of them, at least—took to the rails, living in hobo jungles, perhaps looking for a way to earn a few dollars, perhaps just trying to distract themselves from hopelessness in the grand old American way: movement, going here and there and everywhere for the mere sake of the journey, fearing arrival at a cold new destination as much as their ongoing poverty.

By 1933, the unemployment rate in the United States, which had been about 3 percent eight years earlier, had soared to 25 percent. An estimated 7,000 people in New York City alone set up their own shoeshine stands, although few were the customers who could afford such a luxury. Others sold apples on the sidewalk, and most days were lucky to make enough money to buy a few pieces of fruit for their own families when they went home. And a grand total of three men found work through the '30s as "butt pickers" at New York's Rockefeller Center, picking up and disposing of the remains of all the cigarettes smoked so nervously in the complex's various buildings.

The times had already been hard for farmers. Prices of agricultural goods fell about 40 percent in 1920 and 1921 and remained low throughout the decade. But tobacco farmers did not hit bottom until the stock market did. In 1928, the strain of leaf known as Burley, among the most popular for cigarettes, sold for thirty-one cents a pound. By 1931, it was going for less than nine cents a pound, and many of the men who grew it could not see a reason to work as hard as they did to lose so much money. Some of them, especially the smaller farmers, gave up, letting existing crops rot and failing to plant new ones. They joined one exodus or another, by foot or freight or automobile, changing their location if not their luck, still wanting to smoke tobacco but never again willing to grow the stuff.

Joseph C. Robert tells the story of a young farmer from New Canton, Virginia, named Dan Wood. He was a "sharecropper with a wife and two children [who] took a load of 414 pounds of dark tobacco, good but not fancy, 30 miles to market. When the carriage charges were deducted, the check amounted to $5.19; his share was $2.60." Both the raising of the tobacco and the journey to dispose of it had cost more.

Cigarette sales took a dive early in the Depression: from 124 billion in 1930 to 117 billion in 1931. In fact, nothing seemed to go right that year for the industry. When one company came out with a new brand called Fems, featuring a red mouthpiece that hid the mark of a woman's lipstick, women did not care; the company was out of business within months. This was not a time for gimmicks. A few other companies ceased to exist, as well. For the most part, they were the newer, less well-established firms, and they had no money to advertise to people who had so little money to buy.

But as sales were sinking in '31, ad budgets for the established firms were shooting up. The tobacco industry spent 75 million dollars to entice smokers that year, the most ever, and sales climbed in 1932 and continued to climb, although in gradual fashion, for the rest of the hard times. So, even more gradually, did income for tobacco farmers. Shortly after Black Tuesday, the average American family was allocating about 4 percent of its available money to tobacco. In the 1930s, the figure went up, in some cases to as much as 7 percent. Somehow, men and women found a way to find the money to take their minds off their poverty, at least for a few minutes, a few blessed puffs.

Still, leaf profits were not what they used to be. Some people lit up as much as before but smoked only half a cigarette at a time. Others smoked fewer packs than they once had, one a day instead of two, or ten a week instead of fifteen. Still others did not smoke packs at all; they bought "loosies," individual cigarettes that were sold on the street for a penny apiece. Since the average pack cost fifteen cents and contained twenty smokes, buying "loosies" in bulk would have made no sense. Those who smoked them, however, could no longer afford to indulge their tastes in bulk, as much as they might have wanted to.

And then there were men and women who did not buy ready-made cigarettes at all, instead purchasing cheap tobacco and cheap papers and rolling their own. The cigarettes did not usually turn out well; they were hard on both the draw and the taste buds and sometimes fell apart in a person's mouth. But a hand-roller saved a few cents here, a few there; it was everyone's goal during the Depression.

Another good, cheap pleasure of the time was movies. Men and women went to theaters like pilgrims to shrines and, sitting before big, glowing screens, were transported to other places as surely as if the theater were one of those hobo-laden boxcars rattling across the country. They reveled in the glamour that they saw and laughed at the misadventures and bit their nails at the suspense—so ripe were they for stories not their own, lives that seemed more purposeful and hopeful. They envied the stars up there on the wall, pretending to be all sorts of interesting characters, so many of them treating cigarettes as an accessory to their charm and seductiveness and daring. But it was a benign form of envy, a longing more than resentment. That someone could be a fabulously wealthy, fabulously attired, fabulously venerated actor or actress during these debilitating years offered hope to all—or, at least, hinted that the despair which seemed so pervasive would not last forever.

And if it did not, as Iain Gately writes, if they "could not afford the mansions, the yachts, the furs or the diamond their idols enjoyed in flickering black and white ... they could buy the cigarettes and so share a portion of the dream. Smoking was an aspiration everybody could fulfill."

In fact, it may be that when moviegoers walked out of their theaters, blinking at both the renewed brightness and the descent back into harsh reality, they wanted a smoke even more than they did when they went in:

> Early in the depression several companies, led by Warner Brothers, released films of social conscience in which the criminal or member of the lower class, portrayed by actors such as James Cagney, Paul Muni, Humphrey Bogart, and George Raft, smoked cigarettes, even while gunning down opponents. Given the quasi-revolutionary atmosphere of the time, these men were looked upon as heroes. After several protests made their way to Hollywood and the studios had time to react, the same actors were cast as FBI agents or policemen, but the cigarettes remained. Working women, often their consorts or wives, played by Ann Dvorak, Ann Sheridan, Bette Davis, and others, also smoked cigarettes. ... In this way, the most important and effective entertainment medium of the 1930s served the interests of the cigarette far better than did the paid ads put out by the Madison Avenue copywriters.

The interests of pipes and cigars were served on screen, as well, during the Depression, with the man who smoked the former being portrayed as a thoughtful fellow, solid in his values, dependable in his behavior. Those who partook of the latter on celluloid were more often than not figures of power, tycoons, high-ranking officials in government, sometimes even mob bosses. And the cigarette smoker could be any number of types, could reveal any number of traits:

> The way a man or woman handled a cigarette on screen betrayed not only their background, but also their state of mind. Fingers could shake, matches spill to portray nerves, a cigarette could be lit effortlessly, flamboyantly (literally), sexily, intriguingly—the possibilities were numerous, and Hollywood exploited them all.

On an annual basis, Americans smoked more during the Great Depression than they did during World War I; this, after all, was a peace riddled to the core with wartime anxieties. And they did so without worry about their health, about fouling their lungs or infecting their sinuses or predisposing themselves to any number of disabling, if not fatal, ailments. They would never be able to smoke in such innocence again. The worst of all years for the American as breadwinner would be the last good time for him, and her, as smoker.

TEN

The Case against Tobacco

JAMES I WAS RIGHT. Not specific or fair-minded, not politic, and certainly not civil to those who took issue with either his conclusions or his tone. But somehow the man knew what he was talking about in his *Counterblaste to Tobacco* when he agreed with Philaretes that the weed "makes a kitchen of the inward parts of men, soiling and infecting them with an unctuous and oily kind of soot." Elsewhere in the document he wrote of danger to the lungs, which he did not think could function properly if exposed to hot, foul-smelling smoke.

But there was no way to prove these statements in the seventeenth century, in England or anywhere else. James seemed to be expressing an opinion, nothing more, and an unpopular one at that. Only later, much later, would people learn that his rantings—based, it seemed to some, on nothing more then baseless ill will—were actually statements of fact.

The French, though, suspected as much only a few decades after the *Counterblaste.* In 1631, their Parliament voted to prohibit the leaf in all forms for people serving terms in prison, this after doctors announced that inmates were among the least healthy persons in the land, and tobacco seemed to be at fault. Precisely how they did not say, but their certainty led to perhaps the first ban ever imposed on smoking for reasons of health by a government that did not threaten violence for disobedience.

According to the journalist Susan Wagner, whose chronology is as good as any, "The scientific study of tobacco and its effects on the body may be said to have begun in 1671, when the Italian biologist Francesco Redi published an account of the lethal effects of the 'oil of tobacco.'" But to call Redi's account scientific is to look at it from the vantage point of the present. At the time, he, too, was opining more than proving, and nobody wanted to listen, much less believe.

About eighty years later came another disregarded theory, this one from an institution rather than a mere individual. The Medical School of Paris declared that using the weed shortened a person's life. Not that it *might* shorten a person's life; that it *did*. But the school provided no more detail than did the French doctors who had looked into the prisons, at least not the kind of detail that average Frenchmen could understand or the more learned among their countrymen could verify. Besides, medicine was not in those days the august field it has since become. Its cure rate, for almost all diseases, was more likely to inspire trepidation than confidence. The Medical School's findings were met with an initial burst of indifference, and then a far lengthier one.

In the early nineteenth century, at about the time that Benjamin Rush was trying to impress upon Americans the hazards of tobacco, a French chemist named Louis Nicholas Vaquelin discovered the leaf's principal active ingredient. He named it after one of his own countrymen, the former ambassador to Portugal from the court of Catherine de Medici, Jean Nicot. It is not clear how dangerous Vaquelin thought nicotine was, or whether he thought it was dangerous at all. "Active," he called it, and nothing else.

By the middle of the nineteenth century, the study of human bodily malfunctions had made important advances in some ways while continuing to reside at the level of superstition in others. It was, in one author's opinion, "an age when the best that medical science could offer seemed to be a choice between bleeding and port wine." There were still more expressions of anxiety than of confidence and probably more benefits available from the port than from any of the era's nominal medicines.

Which may explain what happened in 1857, when the British medical journal *Lancet* became the most reputable source yet to make specific charges against tobacco. *Lancet* reported on research showing that the leaf slowed the workings of the mind by producing drowsiness and rendered the mind less efficient by creating irritability. The article went on to say that tobacco caused damage to the respiratory system, as well as to the larynx, trachea, and bronchae. It definitely made people cough, as had long been suspected, and almost surely had an adverse effect on the heart's ability to circulate blood. In the long term, the research suggested, tobacco's effects could be very harmful.

What *did* happen in the wake of the article? Nothing. Not a single recorded gasp or misgiving or second thought. The British citizenry could not have been less alarmed by the *Lancet* conclusions, could not have registered them to a less perceptible degree, if they had been announced in a foreign language or in invisible ink. The popular press does not seem to have reprinted any excerpts from the findings or to have offered a summary, and those people who actually knew of them found it easier to believe *Lancet* was wrong than that their favorite form of flammable vegetation was a risk to health and longevity. The Brits kept on smoking, even picked up the pace a little, and *Lancet*'s charges never made it to the United States.

Two years later, back in France, a doctor named Bouisson surveyed sixty-eight patients at a hospital in Montpelier. All had cancer of the mouth, tonsils, tongue, or lips. All were smokers. Sixty-six of them took their tobacco through short-stemmed clay pipes. Seldom, in Bouisson's experience, had cause and effect seemed so clearly established.

Excitedly, he alerted the nation. Calmly, the nation reacted. Companies that made short-stemmed clay pipes found there was less of a market for them than before, so they began manufacturing long-stemmed clay pipes, which people bought avidly. Bouisson could do little but shake his head.

But inroads were being made, ideas beginning to form, however indistinctly, however unpopular the initial receptions. In 1879, the *New York Times* referred, again without proof, to "the disastrous effects of nicotine upon the human system." Other newspapers began to run similar stories, making serious charges in general terms. Magazines followed suit, but in most cases only once. As yet, there was simply not enough to say for multi-part articles or follow-up reports.

And fiction was coming around as well as fact. Eminent authors had glorified tobacco in the past; now a few would begin to warn against it. They did not do so often, and they were tentative at first. Furthermore, the characters who doubted the weed on their creators' behalf were never the most admirable or virile in the books that housed them.

Still, there is a noteworthy passage in Henry James's classic 1881 novel *The Portrait of a Lady*, in which Ralph Touchett, a dapper if somewhat limited gentleman, is found to be "thoroughly ill. He has been getting worse every year, and now he has no strength left." So what does he do? "He smokes no more cigarettes!" It was, for the period, a drastic step. Not to mention an ineffective one; Touchett died anyhow, though most likely from other causes.

As the nineteenth century drew to a close and the twentieth took its first few tentative steps, other responsible agencies either verified the *Lancet* discoveries or made similar ones of their own. And in 1912, an American physician named Tidswell followed up on M. Bouisson's work in building the statistical case against the leaf. He might have been a bit sweeping, though, when he suggested that "the most common cause of female sterility is the abuse of tobacco by males. ... [T]hose countries which use the most tobacco have the largest number of stillbirths." The charge was repeated in other periodicals in the next few years, and a variety of defects in surviving infants were also attributed to the weed.

But it is probably accurate to say that not until the 1920s, when so many Americans were smoking with such pride and abandon, and when the tobacco companies were spending more on their ad budgets than most Central American countries were spending on the sum total of their goods and services, did construction began on the final case against tobacco, and then only in preliminary form.

1921: Dr. Moses Barron of the University of Minnesota notices something peculiar. Examining autopsy records from 1899 to 1919, he discovers only four cases of lung cancer at the university's hospital. In the twelve-month period ending on June 20, 1921, Barron finds eight cases. He suspects that tobacco is responsible for the increase, as people are smoking more now than they did before, but the statistical sample is too small for him to be certain. He says nothing, but his investigations continue, and other researchers take note.

1924, England: Sir Ernest L. Kennaway, a chemist of some standing, publishes a paper on an ingredient in tobacco known as tar. It is a greasy substance, dark brown or black, produced by the solids in smoke when they settle. Kennaway is repulsed by it. He tests it, analyzes his results carefully. He concludes that the substance causes cancer, and it is a thunderbolt of an accusation.

1925: Kennaway keeps on with his work. Now he tells journalists that he is "able to produce cancerous tumors on the skins of laboratory animals by painting them with condensed smoke 'tar.'"

The experiment is the subject of much debate in the medical and scientific communities. But the tobacco community, when it deigns to acknowledge Kennaway at all, does so condescendingly. It points out that what he has done with his little rat-size paintbrush is something quite different from inhaling and exhaling the smoke of a cigarette. This seems to most people a satisfactory response; they will continue to light up, but under no circumstances will they slather their bodies with brushloads of condensed tar, even if they can somehow get their hands on the stuff.

1928, United States: Drs. Herbert L. Lombard and Carl R. Doering announce the results of a study they have conducted of 434 men and women in Massachusetts, half of whom have cancer, half of whom do not. Among other things, the doctors deduce that heavy smokers are 27 percent more likely to contract the disease than non-smokers, and the more a person smokes, the more the odds increase.

The cigarette industry objects to the statistics. In fact, it objects to the whole notion of numbers as a means of proving causation. This is specious, the industry says, unfair. A variety of other factors might be responsible for the cancer, factors that would certainly show up if they were the focus of the research rather than a tobacco product.

1929: Scientists in Cologne, Germany, undaunted, make the strongest statistical correlation yet between smoking and cancer—so strong, they insist, that causation is certain. Shortly thereafter in the United States, Drs. Alton Ochsner and Michael De Bakey publish findings of their own that confirm the link. When challenged, Ochsner and De Bakey claim that they have studied too many cases, found too many links. Neither coincidence nor other primary causes are possible.

1938: Dr. Raymond Pearl of Johns Hopkins University reports to the New York Academy of Medicine on "The Search for Longevity," reading his paper at the group's convention. "Smoking is associated with a definite impairment of longevity," he says. "This impairment is proportional to the habitual amount of tobacco usage by smoking, being great for heavy smokers and less for moderate smokers." Pearl studied 6,813 men and women; "two-thirds of the non-smokers had lived beyond sixty; 61 per cent of the moderate smokers had reached the same age; but only 46 per cent of the heavy smokers reached age sixty."

1934–38: As more Americans smoke, more die from lung cancer. According to the U.S. Bureau of the Census, there is a 36 percent increase in such deaths during this four-year period.

1939: In Germany again, Dr. Franz Muller "presents the world's first controlled epidemiological study of the tobacco–lung cancer relationship." Muller acknowledges "the usual list of causes—road dust and road tar, automobile exhaust, trauma, TB, influenza, X-rays, and industrial pollutants—but argues that 'the significance of tobacco smoke has been pushed more and more into the foreground.'"

To find out just how far into the foreground, Muller devised a survey that he sent to relatives of victims of the disease. Among the questions:

1. Was the deceased, Herr ________, a smoker? If so, what was his daily consumption of cigars, cigarettes, or pipe tobacco? (Please be numerically precise in your answer!)

2. Did the deceased smoke at some point in his life and then stop? Until when did he smoke? If he did smoke, what was his daily consumption of cigars, cigarettes or pipe tobacco? (Please be precise!)
3. Did the deceased ever cut down on his smoking? How high was his daily use of tobacco products, before and after he cut back? (Please be precise!)

Meanwhile, another German scientist, Dr. Fritz Lickint, publishes *Tabak und Organismus*, "arguably the most comprehensive scholarly indictment of tobacco ever published." It blames the weed for cancers from one end of the "smoke alley" to the other: "lips, tongue, lining of the mouth, jaw, esophagus, windpipe and lungs." Lickint is also the first to notice that there seems to be a link between what will come to be called "secondhand smoke" and illness.

Both Muller and Lickint are in the service of the weed-hating Adolf Hitler, who has already forbidden pilots of the Luftwaffe to smoke and intends a worldwide ban of tobacco products once he conquers enough of the world to make the plan feasible.

BUT THE EFFECT of all these data was not what reformers had hoped. It did not persuade smokers to give up their habit so much as it motivated people already opposed to the weed to renew their opposition and to get more organized and dedicated about it. For the first time, the opponents had what seemed to be solid and dispassionate evidence on their side; facts now supported what earlier had been too easy to dismiss as prejudice or single-mindedness. There were still hurdles to overcome, though, still a centuries-old tradition of acceptance of tobacco, still a fascination for smoke and the wispy powers it seemed to possess.

The Anti-Cigarette League no longer existed, and with Prohibition having been repealed after establishing itself as perhaps the greatest legislative debacle in American history, there was no impetus for a well-organized, nationally based movement to take the league's place. But as individuals and in small groups, at universities and in other outposts of research, a new generation of tobacco foes was picking up the threads of their predecessors' work. They were pressuring lawmakers, writing letters to newspapers, lecturing family and friends and students and the members of various professional associations, seeing to it that the results of the latest medical studies were made available to all, and urging even more studies, certain of more damning conclusions. Alcohol might have been legitimized—or re-legitimized—by the repeal of the Eighteenth Amendment; the tobacco reformers could not let the same thing happen to the leaf.

In the past, cigarette companies would have ignored the opposition. The weed haters were too few, their credentials too unimpressive, and their accusations too ineffective to deserve replies. But the new charges, originating as they did in laboratories rather than in the imaginations of zealots, were formidable. The cigarette industry would have to go public, say something on its own behalf. The problem was, what? Exculpatory data would have been nice, but since the companies were wholly lacking in it, they decided to fight fire with money, of which they had plenty. They used the money to create advertising slogans and design publicity campaigns that addressed the charges against cigarettes with Brobdingnagian duplicity:

Old Gold

"Not A Cough in A Carload."

Camel

"Not a Single Case of Throat Irritation Due to Smoking Camels."

"For digestion's sake, smoke Camels ... stimulates the flow of digestive fluids."

"More Doctors Smoke Camels Than Any Other Cigarette."

Fatima

"... truly comfortable to your throat and tongue."

Kool

"Guards Against Colds."

"Doctors ... lawyers ... merchants, chiefs in every walk of life agree that Kools are soothing to your throat."

Philip Morris

"The Throat-Tested Cigarette."

"Many Leading Nose and Throat Specialists Suggest ... Change to Philip Morris."

Among those who endorsed Lucky Strike was the opera singer Ernestine Schumann-Heink, one of the few in that field to become a favorite of the mass audience, a female Mario Lanza or Luciano Pavarotti well ahead of her time. According to an ad that Lucky ran in a number of publications, the diva owed at least some of her success as a vocalist to George Washington Hill's favorite product, for not only did the smoke not impair her performance, it made her sound better, resonant and ringing, the mistress of her octaves—something about the effect of the smoke twisting and curling so sweetly around the vocal chords. Without Lucky, she would be just another belter; that, at least, is what readers were led to think.

The ad seemed a successful one with the general audience. But it was a flop on the college campuses at which Madame Schumann-Heink had been scheduled to perform. The schools, having been influenced by anti-smoking reports and, perhaps in some cases, by research being done on those very campuses, canceled her appearances. One day the singer had a fully booked concert tour ahead of her, the next a series of empty and non-remunerative nights at home.

Madame Schumann-Heink was distraught. She hastily reconsidered her position. She said that her signature, which was reproduced at the bottom of the Lucky ad, had been obtained under false pretenses, although she did not explain how. She said she did not smoke, had never smoked, did not allow her sons or anyone else to smoke in her presence. The use of the weed was an abomination to her. She was truly sorry, and more than a little mystified, that her position had been so grossly misrepresented by the manufacturers of Lucky Strike. It just went to show how far the forces of tobacco would go to make a point.

The story worked. Madame Schumann-Heink got her gigs back—most of them, at any rate—and demonstrated her sincerity, freshly minted though it seems to have been, by continuing to blast cigarettes when she performed on campus. She had learned a lesson she would never forget. She had learned that the times were changing.

But in the majority of cases, the fanciful claims of the tobacco companies went unchallenged. People enjoyed their smokes too much to yield to nay-sayers, especially since, relatively speaking, there were so few of them. Yes, they were doctors and scientists and others with advanced degrees, but no one had ever heard of them before. Who *were* these guys? Who knew about their credentials, their motives? There were frauds in every line of work, even the most prestigious ones. Smokers could not figure out what to make of such killjoys, as sober and insistent as deacons in their pronouncements of doom. And so, rather than heeding them, they continued to rely on their own instincts.

And they relied on the examples of their heroes. For instance, there was Lou Gehrig, the New York Yankees' first baseman, who swore in an advertisement that "Camels don't get your wind." There was Joe DiMaggio, the same team's centerfielder and another of the all-time baseball greats, who smoked not only in his private hours but during the game. DiMaggio biographer Richard Ben Cramer tells of another Yankee outfielder, a fellow named Hank Workman, who seldom got into a game. "But he had another job: as each inning ended, he had to light a Chesterfield, take one puff and have it burning for the Dago [DiMaggio] when he came in from center field." And there were other athletes, and actors and singers and comedians, men and women who appeared both in advertisements and in public places, sucking on their cigarettes, smiling their smiles, showing the way for their fans.

Earlier, in 1909, Honus Wagner, the Pittsburgh Pirates' Hall of Fame shortstop, had demanded that the American Tobacco Company remove his picture from its packs of Sweet Caporals. He had begun to think that children were starting to smoke because of it, and Wagner did want such a thing on his conscience. But by the 1930s and '40s, few athletes remembered Wagner's example, and fewer still cared.

And there was more. The weed had history on its side. It had saved Jamestown and perhaps the entire process of British colonization in the southeastern United States. It had relieved the American soldier in World War I. In between, it had given countless hours of satisfaction to countless numbers of men, women, and, yes, children. To some it was a psychological crutch, to some a favorite pastime, to some an irresistible taste, to some a statement of manliness or womanliness, to some a means of occupying the hands in a social setting, to some a means of fitting into that social setting with a minimum of discomfort. People thought of tobacco as they thought of metals or grains, livestock or textiles—an important part of the nation's economy, a building block of society, something as dependable as it was valuable. It was a necessity, not a luxury; a blessing, not a curse.

In 1911, the average American smoked 141 cigarettes. In 1941, he smoked 1,892, almost fourteen times as many.

Also in 1941—on December 7, to be exact—the Japanese air force attacked the U.S. military installation at Pearl Harbor, Hawaii, sinking eight battleships, three light cruisers, three destroyers, and four other naval vessels, as well as crippling 170 U.S. fighter planes on the ground. It seemed that the tobacco reformers, whoever they happened to be and whatever they were now able to prove, could just not get a break.

IT WAS PRESIDENT FRANKLIN ROOSEVELT, so often photographed with that cigarette holder of his, smiling and confident, holding his chin up and out, jaunty as could be, who first spoke to the importance of tobacco in World War II. He proclaimed the leaf an essential crop and ordered draft boards to defer the men who grew it whenever possible. They responded as the rest of America's labor force responded to the challenge of the Axis powers—by working longer hours with greater efficiency than they had ever worked in the past and, as a result, boosting production, an 18 percent increase in tobacco yields between 1940 and the armistice in 1945. It was one of the biggest jumps ever recorded for any agricultural commodity in so short a time, and the farmers were well compensated for it. The price they charged for their crop during the war years went from 16.1 cents per pound to 42.6. In World War I, though, the percentage of increase had been even higher: 9.7 cents to 31.2.

The tobacco companies also made money. Such was the demand for their product that they did not have to provide it free anymore, although they did offer substantial discounts to the troops. Several firms charged about a third of the usual rate, and Lorillard accomplished the same thing by allowing GIs to buy three cartons of Old Gold for the price of one. But the sales volume was so great that the companies still showed a profit, especially when the government decided to remove taxes on all merchandise, tobacco or otherwise, with a military destination. During World War II, almost one out of every five cigarettes made in the United States was smoked by an American fighting man, either in training or at the front.

And no one seemed more grateful for the soldiers' bounty than General Douglas MacArthur, Supreme Commander, Allied Forces, Southwest Pacific. A citizen's group that had raised ten thousand dollars for his men asked the general what to do with it. The group had in mind blankets or beverages or perhaps inspirational literature of some sort. As had Washington and Pershing before him, MacArthur set the civilians straight. "The entire amount," he said, "should be used to buy American cigarettes which, of all personal comforts, are the most difficult to obtain here." Although a pipe smoker himself, and a frequent one, MacArthur believed that cigarettes made his troops, if not more competent warriors, at least more contented ones. Cigarettes, in other words, took some of the burden off the general.

But MacArthur had a problem. His men had a problem. They were stationed in the South Pacific, and the South Pacific was one of the most humid places on earth to fight a war—or do anything else, for that matter. Cigarettes arrived there in smokable condition, then within hours became so soggy that there were times when they could not even be lit. Even when they could, the draw was often difficult and the taste not up to normal standards. MacArthur's troops were disappointed and quick to complain.

Fortunately for them, the firm of Larus & Brothers had heard about the situation and already figured out a way to help. Rather than keep on shipping packs of its Chelsea brand to the war zone in their usual cardboard boxes, the firm packed the packs in cans—Planters' Peanuts cans. Within a single year, Larus & Brothers had shipped 7 million peanut cans of cigarettes to our boys in the damp, dank Pacific, a great many of whom were so appreciative that they not only smoked every Chelsea they could get their hands on during combat but remained faithful Chelsea smokers until Larus & Brothers went out of business many years later. Some of them claimed that they went so far as to develop a fondness for peanuts, even though they had never cared for nuts of any kind before they went off to war in the tropics. And when they would see the cans in grocery stores in the years ahead, they would smile, thinking of them fondly, as little vaults at a time of crisis.

Other cigarette companies capitalized on the fighting in their own ways, as a new generation of ad slogans reflected the nation's new reality:

Chesterfield

"Keep 'em Smoking, Our Fighting Men Rate the Best."

Camel

"Camels are the favorite! In the Army ... In the Navy ... In the Marine Corps ... In the Coast Guard!"

"You Want Steady Nerves When You're Flying Uncle Sam's Bombers Across the Ocean."

Lucky Strike

"Lucky Strike Green Has Gone to War"

The Lucky Strike slogan, despite seeming the most innocuous of the batch, was probably the most effective, and therein lies another Lucky tale.

It seems that, despite Edward L. Bernays's best efforts to the contrary, the Lucky Strike pack remained a problem. Tests continued to show that green was an off-putting color, not just for women whose fashion sense had turned away from it again after the big green blitz of the '20s, but for men, who simply indicated to pollsters, without giving a reason, that the shade had no appeal. George Washington Hill did not understand why, but in analyzing his competitors' products, he came to believe that the Lucky pack "lacked the 'cleanliness' of the white Camel and Chesterfield wrappers, or the richness of the warm brown of Philip Morris." Something had to be done, and this time Hill did it himself, deciding, for some reason, that Bernays was simply not up to the challenge.

With the outbreak of World War II, Lucky Strike went white. As Robert Sobel explains, more than just color was involved in the decision:

> The implication was that American Tobacco had stopped using the color [green] so as to make the ink available to the armed forces, perhaps for camouflage paint. In fact this was not the case; Lucky Strike green was not needed by any branch of the defense effort, but in terms of advertising impact it was a brilliant idea. Camel and Chesterfield might employ military themes in their ads, Hill seemed to be saying, but Luckies was making the real sacrifice.

American soldiers were even more obsessed with cigarettes in World War II than they had been in other wars. After all, more of them had been smokers when the war started, both in absolute numbers and as a percentage of the population. In a few cases, the obsession led to undue apprehension; sometimes running out of tobacco seemed a bigger concern to GIs than running out of ammunition. In Virginia, an army base newspaper called the *Camp Lee Traveler* reported on a self-fulfilling dread: "The soldiers in worrying about a possible shortage were smoking more than usual and thus helping to create the very situation they were fearing."

Dr. Earnest Albert Hooton, a Rhodes Scholar turned Harvard anthropologist, understood their anxieties. He also understood why so many young ladies at home were similarly anxious:

> The boys in the fox-holes, with their lives endangered, are nervous and miserable and want girls. Since they can't have them, they smoke cigarettes. The girls at home, with their virtue not endangered, are nervous and miserable and want boys. Since they can't have them, they too smoke cigarettes.

But it was not just girls. Everyone on the home front seemed to be smoking, people of all wartime occupations and ages. Adolescents in particular, with their fathers overseas and their mothers filling in at work, were now left unsupervised to cultivate almost any vice of their choosing. Many of them chose the patriotic one. As a result, per capita consumption of small smokes in the United States increased by a third during the years of combat, from 1,551 per person to 2,027. Consider the context: first the tensions of World War I, then the breakneck pace of the '20s, then the haunting penury of the Great Depression, now the greater tensions of a war in which the United States had a more personal stake than it did a couple of decades earlier, as well as greater obstacles to overcome and a minimum of time to do it. To Americans of a certain age, the early '40s were the most nerve-wracking of all possible times, and the cigarette the most accessible of remedies.

The Camp Lee experience notwithstanding, there were few shortages of tobacco in the military. All branches of service realized the importance of the weed to their men and women; all branches kept their shelves fully stocked and their lines of supply open. Cigarettes were usually a part of a soldier's daily ration, in most cases between five and seven packs a week, with additional quantities available at the PX, which never seemed to run out.

It's a good thing they didn't. "From the soldiers' point of view," it has been written about World War II, "cigarettes were more than mild anaesthetics. Cigarettes formed an umbilical cord linking soldier to civilization. There was little else in the daily grind of being bombed, burned and maimed, of killing or being killed in foreign countries to remind them of home."

Civilian life, on the other hand, was plagued with shortages. In many cities—Chicago, Atlanta, Detroit, and Philadelphia among them—men and women found out when their favorite stores would be getting shipments of cigarettes and lined up hours in advance, sometimes even the previous day. They were like fans who would camp out later in the century for tickets to the Super Bowl or a Bruce Springsteen concert. And the longer the lines, the shorter the tempers of those who formed them. It was not uncommon for people to bump into one another and for the bumping to turn to shoving and even fist swinging. On occasion, people pulled weapons; other times they swore revenge for some point in the future, after they had gotten their smokes and fortified themselves with a few drags. Police were called from time to time, and arrests were made. There was too much at stake in the cigarette lines for shows of conventional civility.

Newspapers of the time told of scalping, with opportunists at the heads of the lines buying as many cartons of smokes as they could and then selling them to people at the ends of the lines for ten, twenty, or even thirty times what they were worth, whatever they could get. There were no laws against it. Other reports told of people puffing away to kill the time in line, and persons standing near them getting so jealous that, like Lucy Page Gaston's "stinkers," they snatched the cigarettes from their neighbors' mouths and ran away, inhaling frantically as they tried to escape. The victims faced the choice of either pursuing their tormenters and losing their places, or losing their smokes and staying put to buy more.

In Chicago, where certain kinds of criminal behavior had become almost institutionalized because of Prohibition and the newspaper circulation battles that preceded it, tobacco delivery vehicles were sometimes hijacked on the roads or raided in front of stores as they unloaded. The thieves converted the cigarettes to cash on the black market, charging inflated prices to people who could afford not to wait in line or who could not get the brands of leaf they wanted any other way.

Eventually, a few tobacco companies began to deliver their cigarettes in armored cars. They also hired extra guards and varied their routes and delivery times, taking all possible precautions, so valuable was the cargo, such security did it require at a time of scarcity. New brands, many of which would soon be defunct, like Wings and Home Run, came to market to meet the swollen demand; old brands pleaded with their customers to remain faithful. "You can't always get them," went a slogan of the shortage days, "but Camels are worth asking for."

Yet for all the smoking and all the hectic striving by people to get their fixes, something was different about attitudes during World War II, at least on the home front. Americans seemed to know they were taking a chance, that the momentary pleasure they were experiencing with their tobacco might have long-term costs, that disease and maybe even premature death might await them for their appetites. There had been so many warnings from doctors and scientists, so many reports and statements and studies. Cigarette smokers could not help but wonder about them. And they could not help but wonder what new doubts would be cast, what new alarms would be sounded, and how loudly and how soon.

WHEN THE WAR ended, after General MacArthur had offered the emperor of Japan a cigarette when receiving his nation's formal surrender and noted "how his hands shook as I lighted it for him," and as congressional debate began on the Marshall Plan, which would eventually, among other things, provide more than 200 million cigarettes to help promote the rebuilding of Germany—in other words, after peace had returned to a world so recently bellicose, Americans tried to get back to their old routines as fast and as optimistically as possible. But the preceding years had changed them, and as they were now different people, so would they lead different lives.

One reason was that they had new and more profound forms of apprehension to cope with, some related to the bomb and the sudden fragility of existence that it appeared to create, others to the war's equally devastating effects on values, ranging from the religious to the vocational, which is to say that God felt uncommonly distant to many Americans, and the purposes of conventional, gray-flannel employment had become increasingly unclear. Some people, returning soldiers and troubled civilians alike, found it harder to satisfy a spouse than it used to be, harder to understand a child, harder to start and maintain a friendship, even to know one's own self. Peace had been won abroad, in the Pacific and in Europe, but turmoil continued at home for a lot of the victors.

However, not so much as a single skirmish of the war had taken place within U.S. borders, so the old structures of life, both physical and cultural, were still standing, up and running as before. The country looked as it always had, and that made it more welcoming than ever to those who had been away for so long and gone through so much in their absence. Not only was the fighting over, so was the Great Depression, and it appeared that the pendulum was beginning to swing all the way over to an exhilarating prosperity, a prosperity in which cigarettes would play an important role. In 1949, according to Richard B. Tennant, the small smokes "accounted for 1.4% of the gross national product at market prices and for 3.5% of all consumer expenditure on nondurable goods."

Furthermore, various people and institutions had dedicated themselves to the specific well-being of the American veteran, with William Levitt and his affordable houses falling into the former category and the federal government, with its GI Bill of Rights providing a free college education and guaranteed loans, into the latter. All things considered, given the extreme nature of what Americans had suffered between 1941 and 1945, both those in the trenches and those at home, it is fair to say that they made a remarkably successful transition from the hellishness of war to the humdrum of civilian life, even if some of them had to smoke too many cigarettes to get over the rough spots.

In Europe it was different. The entire face of the continent had been altered. Thousands of towns and cities and villages were destroyed, and the old ways of life had been reduced in many cases to equally unsalvageable rubble. All the spots, in other words, were rough. Transportation was undependable, communication a struggle. There was not enough food, not enough shelter, not enough in the landscape that was even recognizable to those who had once found the vistas both familiar and reassuring. Death had visited every band of acquaintances, almost every family. One could look at vast stretches of the Old World, could talk to the people who lived there, could share the daily routine of their lives and not know whether they had won the war or lost it. The past was a shadow that extended so far over the present that it obliterated any view of the future.

The measurement of destruction became the grimmest of sciences, and not just in Europe. According to James T. Patterson, writing about the nation that had been punished the most, "World War II had ended with the destruction of 1,700 Russian towns, 31,000 factories, and 100,000 collective farms."

Little was left in the way of either industrial or agricultural capacity, and if there had been a device to show what remained of the human spirit, a Geiger counter of the emotions, it would barely have registered.

A major problem in several countries, Allied and Axis alike, was the collapse of monetary systems. The paper and coins still existed but almost as trinkets; there were few goods and services to back them up, and thus their value fluctuated wildly. The United States did what it could, and to the Soviet Union it issued a set of engraving plates for Allied occupation money, a kind of scrip. But the Russians, too enthusiastic, printed so much of the stuff that it became worthless. They were like children with too many toys, too eager to play.

Other countries could not have afforded to print worthless currency even if they wanted to. The United States was eager to help them, as well, and a number of proposals were considered by both governmental agencies and private concerns. But no one could offer anything in the way of a quick fix.

Ultimately, no one had to. Enter the American cigarette. It was one thing for the weed to have served as legal tender in the place in which it was produced; that had occurred not only in the thirteen original colonies but in other lands, at other times. This, however, was something new: This was one nation's tobacco product being prized so highly that it became an accepted medium of exchange in several other nations, and nothing like it had ever happened before—not, at least, on such a scale.

American soldiers cashed in some of their cigarettes during the war, but not many. There was little to buy in times of actual combat and much reason to smoke. But when the war was over, those still stationed abroad—and there were many of them; the pace of demobilization was grindingly slow—became experts at exploiting the new currency markets. For instance, a fellow would buy two packs of cigarettes for fifteen cents at the PX. Then he would smoke one of the packs and spend the other, buying products he could not have afforded at prewar rates.

One GI remembers that two packs of Lucky or Camel or Old Gold got him a sleeper car from Glasgow to London. Another recalls that three packs paid for a series of tennis lessons in Cannes. A third veteran says that he was able to have an oil painting done of himself for a carton of smokes, although the artist got a little greedy after the deal was made, and the soldier could not pick up his portrait and take it home until he kicked in three cans of sardines in addition.

But neither a carton nor even a whole pack was always required to do business. Ten cigarettes could purchase a good meal in a lot of restaurants, or a bad woman in a lot of seedy neighborhoods. They could purchase groceries and cosmetics and clothing. They could purchase phonograph records and jewelry and fine china, as well as entrée to a fashionable night spot or the use of a car for a day or two or a seat on an airplane that would otherwise have been unavailable. There were even reports of the stubs of cigarettes, the last few fractions of an inch, being accepted as money. People would collect them in a box, like loose change, and offer the box to a street vendor for whatever he happened to be selling: food or clothing, souvenirs or tools. More often than not, the vendor made the deal. Once he had enough stubs, he could rip off the paper and jam the tobacco into hand-rolled cigarettes of his own, which he could either smoke or use for purchases of his own.

Some soldiers, returning home, told friends and reporters that what amazed them most was not how much they could buy for cigarettes but how much respect they were shown just for being able to get hold of them as easily as they did. It was, they said, as if "American tobacco consumer" were a new and pre-eminent social classification. They went on to say that they were often asked whether they had cigarettes not just by people who wanted to trade for them, but by those who merely wanted to identify those who were carrying the treasured articles. It was a form of celebrity worship, of gawking. The celebrity was the person with a pack of Lucky Strike or Camel or Chesterfield in his shirt pocket and needed no more impressive credential for his time or place.

Civilians, too, could capitalize on the lust for cigarettes, and the morality was sometimes murky:

> While Secretary of State James Byrnes and Senator Tom Connally were meeting with the Russians in Berlin in late 1946, Mrs. Byrnes and Mrs. Connally were busily purchasing cigarettes at the PX, and then taking them to the Barter Market, to obtain antiques and other valuables at bargain prices. This was perfectly legal and legitimate. But the following year planeloads of senators and representatives set down in Berlin. Ostensibly, the legislators were there to obtain information for the purpose of framing new measures. But always there was the visit to the Barter Market, cheap cigarettes for works of art and antiques. The Germans saw this and were irate. Herman Goering had plundered the galleries of Europe in the first half of the decade, they said. Now the Americans are trying to do the same in Berlin.

It was not as surprising as it sounds. Cigarettes might have been residents of the vegetable kingdom, objects of longing and loathing, of contentment and controversy, consumer goods and consumer evils—but they were more. They "possessed all the basic qualities for a currency," states historian Robert Sobel.

"They were uniform, easily recognized, universally accepted, almost impossible to counterfeit and, in the beginning at least, had scarcity value as well."

It has been said that "for two years after V-E Day, cigarettes remained the only stable currency in the retail marts of Germany, Italy and France." By the time the two years had passed and European economies had recovered enough to drop tobacco and return to more conventional means of buying and selling, the average American was smoking 128 packs of cigarettes a year. It worked out to one pack every three days, seven cigarettes a day, almost one every two waking hours—a record amount.

In fact, it was twice as much as he had smoked in the flaming, frenzied '20s. He might have smoked differently now, with a "cautious hedonism" rather than unbridled hunger, but he smoked nonetheless and showed no signs of stopping. A cautious hedonism was better than no hedonism at all.

The cigarette industry, of course, was elated, which is another way of saying that it had dropped its guard almost entirely.

WORLD WAR I was the end of the anti-smoking movement that had preceded it. The movement arising in the war's aftermath started virtually from scratch, with new leaders who had new approaches and a different kind of resolve. The leaders were not so memorable as Lucy Page Gaston, but they were much easier to take seriously. The approaches were more rational, less shrill, than before. The resolve was quieter but, perhaps, even more forceful and without question more scientifically based.

World War II, though, was an interruption, nothing more. When it was over, reformers simply picked up where they had left off, building on their earlier advances, adding volumes to the previously compiled evidence. As a result, the 1950s were the worst decade yet for the American tobacco industry.

But it was not just the evidence. No less important than facts and figures, names and dates, were the venues where this information appeared. For the first time, the case against tobacco made the mass media. Charges were now showing up in big-city newspapers and large-circulation magazines, not just esoteric medical journals, and critics could be heard on radio, and occasionally on television, explaining the various dangers of the leaf and backing them up persuasively. It was the establishment turning on the establishment, a period of turmoil for the media as well as for the tobacco companies, since many of the former depended on the latter for a large portion of their advertising income.

The language got simpler, too. The layman no longer had to slog his way through the verbiage of the specialist; he was thus more likely to be frightened by the reports on tobacco's true nature than baffled. And within the language were fewer disclaimers. The case seemed a strong one, nothing equivocal in it, suitable material for the largest possible audiences.

1952: An obscure magazine called the *Christian Herald* publishes an article about research being done by the American Cancer Society on the relationship between smoking and lung cancer. It refers to the research as "a death watch," an extraordinary phrase for the time. It further says that the medical profession is conspiring with "the tobacco industry [to] obscure the truth," which is, the piece goes on to say, that 800,000 young people every year are risking their lives by becoming addicted to the weed.

Few people see the article in the *Christian Herald.* However, it is reprinted in *Reader's Digest,* then as now one of the most read magazines not only in the country but in the world. In part because it accepts no advertising, its objectivity is unquestioned. In the *Christian Herald* the article is called "Smokers Are Getting Scared." The *Digest*'s version is "Cancer by the Carton." Millions of people pore over every word, almost surely the most yet to acquaint themselves with a specific set of complaints against tobacco.

1953: The *Journal of American Geriatrics* states that "the tremendous and unprecedented increase of bronchogenic carcinoma in recent years is due to the carcinogenic effect of tobacco. Because of the likely causal relationship between cigarette smoking and bronchogenic carcinoma, it is our belief that all men who have smoked a package of cigarettes a day for twenty years are likely candidates for bronchogenic carcinoma."

The story is picked up by the Associated Press, International Press, and United Press, the three largest wire services in the country. They distribute it to hundreds upon hundreds of newspapers and radio and TV stations. Almost every one runs it, and when they do, they rewrite the copy to make it punchier, and they refer to the disease as cancer, not carcinoma. Many of the media outlets receive letters and phone calls from readers and listeners wanting to know whether it is really so.

1954: On June 21, in an atmosphere of great anticipation, Drs. E. Cuyler Hammond and Daniel Horn, both affiliated with the American Cancer Society, present a paper to the annual convention of the American Medical Association (AMA) in San Francisco. In it, the authors tell of the exhaustive study they have made of 11,780 men between fifty and sixty-nine, all of whom are smokers. During the period of the study, 7,316 of the men died. If all of them had been non-smokers, say Drs. Hammond and Horn, only 4,651 would have died. Or, to put it another way, the doctors determined this: "The death rate of regular cigarette-smokers was generally 68 percent higher than that of non-smokers, that of smokers of two or more packs a day was 123 percent higher."

The AMA gets more coverage of its convention than it has ever gotten before. "The tobacco industry," Susan Wagner will write several years later, "has never been quite the same since."

1955: The U.S. Bureau of the Census reports that one and a half million Americans have quit smoking in the past eighteen months. As far as anyone knows, it is the greatest number of people who have even *tried* to break the habit, much less succeeded, in so short a time. Press coverage inspires others who have thought about quitting to make the effort, and feature stories, admiring in tone, are written about some of those who succeed.

1957, England: The Ministry of Health, at the urging of the Royal College of Physicians, begins to warn people of the dangers of tobacco. Among other things, the ministry sends out more than a million posters to schools and government offices. They are painfully blunt. One series reads: "Before You Smoke, Think: Cigarettes Cause Lung Cancer." The British also restrict the hours of television commercials for tobacco, pushing them back until late at night, trying to prevent children from seeing them. Similar restrictions are imposed in Denmark and West Germany and actively considered elsewhere.

1957, United States: Our country's Public Health Service is not willing to go so far as the Brits. What it does is make known, in a number of forums, the results of studies indicating that frequent use of tobacco is one of the causes of cancer. Those forums include newspapers, magazines, and addresses by Public Health Service officials. Because this is a first for the federal government—the first position it has ever taken on tobacco—the findings are given column inches by the score and air time by the hour.

1958: The U.S. Public Health Service *does* go further. It says that, after studying 200,000 veterans of World War I and World War II, it has learned that smokers are 32 percent more likely to die at an early age than non-smokers and that heavy smokers are in greater peril than light smokers. The soldiers who had survived one enemy a decade and a half earlier were falling to another enemy now, one whom they had long thought of as a friend.

The story makes the front page of the *New York Times.*

BUT EVEN people who do not read the articles in the '50s know the threat of tobacco. They cannot help but see the headlines as they walk by newsstands; they cannot help but hear people talk about those headlines in the office and at social gatherings; they cannot help but eavesdrop on radio or television discussions from an adjoining room. They cannot help, then, but be apprehensive and, for the first time, begin to consider cutting back, if not abstaining from the leaf altogether.

Concedes Cigarette–Cancer Link
—*Science Digest*, May 1955

Cigarettes: On Trial Again
—*Business Week*, June 11, 1955

Coffin Nails Take Another Beating
—*Christian Century*, September 21, 1955

New Cigarette–Cancer Link
—*Life*, June 11, 1956

It was not like the '20s and '30s. That period was a problem for the tobacco industry. This was a crisis, full-blown and festering and apparent to one and all, a challenge to the industry's very existence. In 1953 and 1954, per capita consumption of the small smoke dropped almost 9 percent, and while the price of many stocks shot to all-time highs during the flush years of the Eisenhower presidency, shares of tobacco firms did not go along for the ride, most instead languishing at prewar—which is to say, late Depression—levels.

The companies responded the only way they could: by increasing advertising budgets. In fact, the tobacco industry spent a higher percentage of its gross earnings on ads than did any other American business in the '50s, or perhaps ever. The single largest expense might have been the industry's "Frank Statement to Cigarette Smokers," appearing in more than four hundred American newspapers, which, it is believed, "reach[ed] a circulation of 43,245,000 people in 258 cities." The statement read, in part, as follows:

> RECENT REPORTS ... have given wide publicity to a theory that cigarette smoking is in some way linked with lung cancer in human beings.
>
> Although conducted by doctors of professional standing, these experiments are not regarded as conclusive in the field of cancer research. ...

> Distinguished authorities point out ... [that] statistics purporting to link cigarette smoking with the disease could apply with equal force to any one of many other aspects of modern life. Indeed the validity of the statistics themselves is questioned by numerous scientists.
>
> We accept an interest in people's health as a basic responsibility, paramount to every other consideration in our business.
>
> We believe the products we make are not injurious to health.
>
> We always have and always will cooperate closely with those whose task it is to safeguard the public health.

In fact, the leaf companies said, they would now cooperate in a new and more positive way by forming something called the Tobacco Industry Research Committee, at the head of which would be "a scientist of unimpeachable integrity and national repute."

But it was not just newspaper pages that the cigarette makers purchased, or minutes on the broadcast media, or even the time and expertise of that scientist of unimpeachable integrity. They also bought themselves entire television shows, and they were very touchy about what the shows portrayed.

On the *Camel News Caravan*, one of the first programs of its kind on television and the immediate predecessor of NBC's *Huntley–Brinkley Report*, anchorman John Cameron Swayze kept a cigarette burning from opening titles to sign-off. He puffed on it during breaks and kept it on the anchor desk, in an ashtray, the rest of the time. As a result, the smoke was often visible as he intoned, wafting around his head and shoulders, filling in the pauses. An oversize picture of a pack of Camels appeared on Swayze's desk; there was no doubt about the smoke's source.

In addition, the program refused to carry stories about the harmful effects of the weed, would not even refer to the latest studies in passing. Neither would it broadcast film that included a no-smoking sign, no matter where or how briefly it might appear. And with the single exception of Winston Churchill, whose cigar had become a symbol of perseverance for the British in World War II and whose sleeping quarters in the Cabinet War Rooms contained a waste basket solely for cigar butts, no smoker was given air time on the *Camel News Caravan*, not so much as a single second or frame, unless it was cigarettes of which he or she partook.

(The War Rooms, it is irresistible to point out, are a remarkable testament to the importance of the weed to our principal ally in the early '40s. One of the most fascinating, yet least known, of modern-day London tourist attractions, they are a series of cubicles of varying size that attach to a dark and meandering basement corridor, with its sole flight of stairs leading up to

St. James's Park. In addition to Churchill's receptacle, there is a bright red box for "Cigarette Ends" in the transatlantic phone room, and on a beam in the map room is a device that looks like a doorbell with a coil next to it. It is, in fact, a specially made cigarette lighter. One presses the "bell" to start the current, then holds one's cigarette to the coil to bring it to life. It was faster than searching through pockets or purse for matches; speed was of the essence in smoking when one was at the same time mapping strategy for war.)

In addition, Camel sponsored an entertainment program called *Man Against Crime*, and it told the show's writers without equivocation that they were not to put a cigarette into the mouth of a criminal or other low-life type, like a philandering husband or a flirtatious wife or the driver of a getaway car. Furthermore, "no one was to be given a cigarette to calm his nerves, since this would imply that nicotine was a narcotic; fire or arson were not to be shown, for this might remind people that cigarettes caused fires; and never, ever, was anyone to cough on a show." The cigarette was to be shown in a positive context, as a fashion accessory or an aid to contemplation or valorous behavior.

Other sponsors were equally inflexible. When a guest on the quiz program *Do You Trust Your Wife?* was asked his mate's astrological sign, he said Cancer. The host thought nothing of it. The producer thought nothing of it. The studio audience, as far as anyone knows, made no untoward associations. But the tobacco barons who paid the bills for the show gulped in resounding unison. Cancer? they said. What the hell kind of sign is *that*? They ordered the episode to be re-shot; in the new version, the spouse was an Aries.

On Madison Avenue, where the admen were growing increasingly nervous about their clients in the leaf trade, and some even considered dropping them for business of a less controversial, more socially responsible nature, the late president of the American Tobacco Company was the subject of a story that rapidly made the rounds, providing a few moments of black-humored relief for harried cigarette sellers. "George Washington Hill would have known what to do about this health business," the joke went. "He would have made cancer fashionable."

But the tobacco firms knew that spending more money on ads and TV programs was not enough. They would have to do something else—something dramatic, attention grabbing, unprecedented; something that showed how much they cared about their customers, that they cared about them even more than they did their bottom lines, and were thus a far more health-conscious segment of the economy than anyone had yet realized, especially those carping cynics in the medical and scientific fields.

So they came up with filters. They brought out cigarettes that had little white tips in front of them and enormous advertising budgets behind them, and, as a result, the industry had a kind of last hurrah, all the American tobacco companies sticking filters onto their smokes and then grinning proudly as they lined up and marched side by side toward the precipice, a mighty glow of ash to lead the way.

ACTUALLY, THERE WAS nothing new about a filter. Dr. Scott's Electric Cigarettes—NO MATCHES REQUIRED; THEY LIGHT ON THE BOX—had had one late in the previous century, and the claim it made was grand. "No Nicotine can be taken into the system while smoking these Cigarettes," read the ad copy, "as in the mouth-piece of each is placed a small wad of absorbent cotton, which strains and eliminates the injurious qualities from the smoke." Of course, hardly anyone believed that there *were* injurious qualities to smoke back then. The market pulled the plug on Dr. Scott's Electric Cigarettes not long after they were introduced, and filters went on a lengthy hiatus.

In fact, not until 1936 did a tobacco company try again, when Kool began to promote a filter made of cork, stating that there was no better way to get rid of impurities—assuming, that is, there really *were* impurities in good old American leaf. More people accepted the premise then; nonetheless, the cork-tipped cigarette found few customers.

But the filters of the '50s were manufactured of far more wondrous materials, of Micronite and Selectrate and Cellulose Acetate and Alpha Cellulose and Activiated Charcoal, to name but a few. One brand had "30,000 filaments." Another had "40,000 filter traps!" Still another boasted that its "Miracle Tip" had been issued U.S. patent number 2,805,671.

Yet for all their differences, the new smokes had something in common. Every one of them made the same promise: A smoker could continue to enjoy all the pleasures of tobacco while at the same time disposing of all the risks. Taste somehow got through the filters; nicotine and tar did not. It was magic of the most convenient kind, a stunning triumph for American know-how—or, at least, American flim-flam.

The slogans that promoted the filter-tips left reality—and in some cases, the rules of grammar—in the dust:

Marlboro

"This filter works good and draws clean."

Hit Parade

"Your Taste Can't Tell the Filter's There."

Winston

"... tastes good like a cigarette should."

L&M

"Just What the Doctor Ordered."

Pall Mall

"... guards against throat scratch."

In 1949, a filter-tipped cigarette was a rarity in the United States. But tobacco companies had been experimenting with the new substances and designs and Unique Selling Propositions for several years, and in bringing filters out in quantity as the '50s got under way, they had the sense of perfect timing. They were right. They also had the sense that they could rehabilitate the entire industry, put the criticism of their foes behind them once and for all. They were wrong. But it would take a while for them to learn this, as the new brands proved an immediate hit with smokers, and their foes, at least for the time being, fell back on their heels.

It was in 1951 that filtered cigarettes were introduced on a large scale, some of them of regular length, others in the new king size. In 1952, they were responsible for a mere 1.5 percent of all cigarette sales. From that point on, their ascent in market share was extraordinary.

1955: 18.7 percent.
1956: 30 percent.
1958: 46 percent.
1960: 50 percent.
1965: 64.4 percent.
1974: 86 percent.
1996: 97 percent.

The most efficacious of the whole bunch of filters seems to have been Kent, it of the Micronite variety, supposedly made of a substance used in hospitals and atomic-energy plants to rid the air of fumes and microscopic particles of dirt. As a result, Kent claimed to reduce by half the amount of tar and nicotine in the average, unfiltered cigarette. Even in an age of hyperbole, to which tobacco had made more than its share of contributions, it seemed an excessive boast.

But it turned out to be true, or close enough to true for cigarette retailing. Lorillard, the maker of Kent, had worked long and hard on its filter and was proud of itself for developing one so proficient. It took out ads to congratulate itself and spread the word among smokers. Executives grinned, patted themselves on the back, and waited for the sales bonanza that would make geniuses, if not millionaires, of them all.

It never came. The Micronite filter, you see, was *too* good. The great accomplishment revealed itself to be an even greater liability, for taking half the tar and nicotine from a cigarette is like taking half the alcohol from a martini; what remains is only a hint of the original taste and satisfaction. The Kent filter was such a powerhouse of a blocking device that customers could barely suck the smoke through it: A little made the passage, a few streams here and there, but it was more of a tickle in the back of the throat than an actual taste in the mouth, and even then a powerful amount of puckering was required. Kents were as much a form of aerobic exercise for the mouth as they were of a nicotine high.

Other filtered brands faced similar problems, and it did not take long, as a result, for consumers to rebel. They thought they had been duped, that the tobacco firms had not only deceived them but provided an inferior product in the process. They grew disenchanted: The more serious the smoker, the greater the disenchantment. And, at least among some smokers, the disenchantment lasted a long time. In his book *Blue Highways*, published in 1982, William Least Heat Moon recounted a conversation he had had with one critic of filters in Greenville, North Carolina:

> Truth is you cain't buy a real, true cigarette anymores. That's why they name them that way—tryin' to convince you what ain't there. Real. True. Nothin' to it. They cut them long, they cut them skinny, they paint them red and green and stuff them with menthol and camphor and eucalyptus. What the hell, they's makin' toys. I'll lay you one of them bright-leaf boys up in Winston-Salem is drawin' up a cigarette you gotta plug in the wall. Nosir, your timber's comin' down to make toys.

In 1957, faced with more than a year of plummeting sales and bad publicity, Lorillard gave in to the prevailing tastes. It told its engineers to restructure the Micronite filter, make it less effective, combine it with a few of the elements of a sieve. This they did, finding the assignment far easier than the original one, and the resulting development went by the name of Micronite II. The ad copy referred to it as new and improved. In reality, it was new and worse. For good measure, Lorillard also doubled the tar and nicotine content of the tobacco that the filter was now not filtering very well. The company had learned its lesson.

Lorillard's competitors followed suit, although since their filters were not as good to begin with they did not have to sabotage them as thoroughly to win back business. But win it back they did, as the flavor returned to the small smokes and excessive amounts of harmful substances returned to the lungs of those who employed them for their pleasure. In 1954, the average American smoked 1,400 cigarettes. In 1963, the total was 4,000.

And that brings us to 1964, the year that had been waiting for the weed like an avenger at roadside ever since the first Maya lit his first pipe all those centuries ago in the jungles of Mexico, ever since the gods were first invoked by smoke and the medicine man first started using the leaf to formulate cures. It was a year that marked the end of all that tobacco had been up to that time in the New World. It would never be the same again.

ELEVEN

The Turning Point

IN THE FIRST MONTH of 1964, Americans were watching *Dr. Kildare* and *Kraft Suspense Theater* and Julia Child, *The French Chef*, on television. They were going to the movies to see Albert Finney in *Tom Jones* and Steve McQueen and Natalie Wood in *Love With the Proper Stranger.* They were listening to "There I've Said It Again," by Bobby Vinton, and "Forget Him," by Bobby Rydell, on the radio, and at hops and canteens and frat parties and night clubs they were dancing to more upbeat songs like "I Want to Hold Your Hand," by the Beatles, and the possibly licentious and undeniably obscure "Louie, Louie," by the Kingsmen.

They were reading Mary McCarthy's *The Group* for titillation and John F. Kennedy's *Profiles in Courage* for inspiration, and possibly solace. The latter had originally been published in 1956 and was a bestseller back then on its own merits. It was an even bigger seller now that the author had been so recently murdered and the desire to keep his memory alive had become a national longing, one of the few ways Americans could think of to deal with a grief so deep and frightening and inexplicable. New York's Idlewild Airport, for instance, had already been renamed for the slain president, and numerous other buildings and highways and bridges and institutions and human infants would adopt the name Kennedy in the weeks and months ahead.

Kennedy's successor, Lyndon Johnson, had appointed the Warren Commission to look into the assassination, and it had already begun its work: collecting evidence, listening to witnesses, evaluating a multitude of theories and surmises and contradictions. Eventually it would release a report that satisfied almost no one at the time and that, even four decades later, would not seem credible to a majority of Americans.

A majority of Panamanians were also upset as 1964 began. They demanded that the canal treaty with the United States be revised in such a way as to acknowledge their nation's sovereignty. When it did not happen, they rioted at displays of the U.S. flag on their soil, demanding that the banners be removed and, in some cases, trying to do so themselves. People were injured, property damaged; at one point, American troops opened fire on the protesters, killing six. The situation seemed at a crisis.

But people in the United States did not take it seriously. In the wake of the Kennedy assassination, it was their own land that gave them pause, not some narrow, humid strip of a country to the south between two oceans. They wondered about conspiracies and uprisings, dark motives and the possibility of corruption in places previously thought to be incorruptible. They were entering a period of cynicism, much of it directed at the federal government, from which they have still not recovered. And they were entering a period of doubt that is still a part of the national temperament.

Insofar as they thought about a foreign nation at all, they thought not about Panama but about South Vietnam. Only a few American soldiers were there as yet, but it had begun to seem that many more would be on the way soon. How else to put an end to the depredations of Viet Cong guerrillas, who were destroying villages; slaughtering men, women, and children; and threatening to topple all of the dominos in that part of the world, to the incalculable detriment of American interests? They *did* have to be stopped, didn't they? They *could* be stopped, couldn't they? The domino analogy *was* valid, wasn't it?

Barefoot in the Park was a big hit on Broadway as 1964 began.

Hello, Dolly! would open later in the month and be even bigger.

The following month, the Beatles would arrive in the United States, causing Elvis Presley-like outbreaks of hysteria and adulation, and some people would take pictures of them at their concerts and other public appearances with a brand new device called an Instamatic camera.

There was a new kind of telephone too, the Trimline, different from previous designs and an easy fit for the hand. You could buy it in colors other than black, even pastels, as if the phone were an item of interior decoration rather than just an appliance. The Trimline had a dial on it, though; the world was still several years away from buttons.

Willie Mays was negotiating a contract with the San Francisco Giants that would pay him $105,000 for the 1964 baseball season—six figures to the left of the decimal point, the fattest deal in the major leagues that year and almost the fattest ever.

On January 11, 1964, fourteen hotels in Atlanta promised they would stop denying admission to people of Mays's race. Some politicians, who had urged the action, called it a landmark, a significant step forward for black Americans. Civil rights leaders, though, were not fooled. They knew that the hotels would not have taken the step, not this early in the century, if they had not been threatened with widespread, business-disrupting, image-jarring protests. There was no reason to believe that there had been, or soon would be, a change of heart, that whites were warming to blacks in the South. There was no real cause for optimism. Still, it was better than *not* being able to get a room or a meal or a drink in the fourteen hotels. Or so the civil rights leaders said to one another, with a certain weary resignation.

And that same day, a Saturday, when a news story of any importance would be sure to get a lot of space in the oversize Sunday newspapers, the nation's highest-ranking medical officer was holding a press conference in Washington, D.C. But would it *be* an important story? Not everyone thought so—at least, not before the event. In the view of journalist Jon Margolis, "few Americans knew that Luther Terry was the surgeon general of the United States. Indeed, not many knew that there was a surgeon general."

The site of the press conference was the auditorium at the U.S. State Department, where President Kennedy had often met with reporters, sometimes overflow crowds of them. In fact, at one such gathering in 1962, after addressing issues of war, peace, and economic growth, Kennedy was asked about the hazards of smoking. He did not seem to have been prepared. He cleared his throat, paused a little too long. "That matter is sensitive enough," he finally replied, "and the stock market is in sufficient difficulty without my giving you an answer which is not based on complete information, which I don't have." Few reporters even mentioned Kennedy's response in their stories.

But now the information *was* complete, or close to it, and the surgeon general had called his press conference to make it available to all. He had chosen a Saturday not just because of Sunday headlines but because the stock market was closed, and would be closed for another day afterward, thereby giving investors time to recuperate from the news, if they needed to, before the opening bell sounded again on Monday.

The early morning hours of January 11 were tense ones for the surgeon general's staff. Many got to work long before they normally did, seeking one another out for support, gathering at tables in the lunchroom and drinking too much coffee from the machine. It tasted even worse on a weekend than it normally did. They watched the second hand on the clock next to the machine, listened to its soft rattle; they talked about subjects in which they had no interest.

J. Stewart Hunter, the surgeon general's assistant for information, expected a big turnout and was not sure how to handle it. "We were scared we'd be stormed by 5,000 folks, all clamoring to get in here." It was for that reason, he later told a reporter, that he smoked fifteen cigarettes that morning before the press conference about the perils of tobacco even started.

He needn't have worried. He certainly needn't have smoked so much. A mere two hundred journalists showed up, not thousands, and they were almost to a person unhappy about it: Working on a Saturday is no reporter's idea of a good time, regardless of the reason. Nor did they seem to think that something historic was about to happen. They believed they had been summoned this morning for a rehash, only that—the presentation of old material in a new package, another attempt to jump-start the smoking scares, which, truth to tell, were beginning to seem less scary to a populace so used to hearing them. The ladies and gentlemen of the fourth estate scattered themselves through the huge auditorium, stretching out in their seats in postures of boredom and impatience. They might have told jokes to the people near them, maybe even the one about smoking being the leading cause of statistics.

What they did not realize was that the very fact of rehashing so *much* information, of adding to it some new facts and figures and interpretations, of arranging the data in a logical and coherent manner, and of releasing it all under the imprimatur of the surgeon general of the United States, making this the first time that the government, still largely respectable and trustworthy as far as most Americans were concerned, had gotten involved in the tobacco controversy to such a degree—what the reporters did not realize was that these factors taken together would give the story far more impact than the apparent sum of its parts, that they would make the press conference something worth covering on a weekend, something, in fact, for the ages.

Both the surgeon general's findings and the account of the Atlanta hotels made the front page of the *New York Times* the next day. But tobacco got bigger headlines than race and more column inches on the inside pages of the first section. Those in favor of a peacefully integrated society still had a long way to go. Those opposed to cigarettes were closing in on their goals.

Actually, the findings were not the surgeon general's alone. Luther Terry had assembled a virtual all-star advisory committee of physicians and scientists and researchers to help him compile and collate all the various materials, make sense of them, and write the report. They came from American's leading medical schools and universities and private research facilities, and since few of these people had ever before taken a public position on tobacco, they were assumed to be impartial.

In fact, three of them, including Terry, were cigarette smokers themselves. Another smoked cigars. Still another did much of his work for the committee with his pipe in his mouth, totaling up the wages of tobacco and puffing away for consolation. Terry also switched to a pipe during the months when the report was being prepared, as he grew more and more alarmed by the evidence against cigarettes. In time, his pipe would alarm him, as well; that, too, he would give up.

The other five panel members were, and would continue to be, nonsmokers.

Before the press conference got under way, Terry's aides passed through the State Department auditorium, distributing copies of the report, two of which had already been sent by special messenger to the White House. It was called *Smoking and Health* and ran to 387 pages and 150,000 words. The journalists groaned at the length. They were told to familiarize themselves with the introduction, which summarized the panel's conclusions, and to skim as much of the rest of the report as they could. They would be given an hour and a half. They groaned at *that* length, too; it was Saturday, after all. But they started in, began perusing, and for the next ninety minutes the only sounds that came from the State Department's largest room were throat clearings and seat squirmings and the flipping of pages and scratching of pens on notepads.

"Cigarette smoking," the reporters read, the importance of the story gradually dawning, "is causally related to lung cancer in men; the magnitude of the effect of cigarette smoking far outweighs all other factors. The data for women, although less extensive, point in the same direction." According to *Smoking and Health*, there had been only 3,000 deaths from lung cancer in the United States in 1930, but the figure had climbed to 18,000 in 1950 and to 41,000 in 1962. The report called this an "extraordinary rise which has not been recorded for cancer of any other site."

But the surgeon general and his committee found tobacco to be culpable in other kinds of cancer, as well, in addition to several non-cancerous ailments. The list was so comprehensive, and the language so direct, that the journalists were not the only people surprised. So, too, were the weed's foes.

Cancer of the larynx: "Evaluation of the evidence leads to the judgment that cigarette smoking is a significant factor in the causation of laryngeal cancer in males."

Oral cancer: "The causal relationship of the smoking of pipes to the development of cancer appears to be established."

Cancer of the esophagus: "The evidence on the tobacco–esophageal cancer relationship supports the belief that an association exists."

Tobacco amblyopia: "Tobacco amblyopia (dimness of vision unexplained by organic lesion) has been related to pipe and cigar smoking by clinical impressions."

Respiratory ailments: "Cigarette smoking is the most important of the causes of chronic bronchitis in the United States."

Peptic ulcers: "Epidemiological studies indicate an association between cigarette smoking and peptic ulcer which is greater for gastric than for duodenal ulcer."

Even cirrhosis of the liver, normally thought to be the drinker's bane, found a place in the surgeon general's report: "Increased mortality of smokers from cirrhosis of the liver has been shown in prospective studies."

Although the report did not link tobacco use to heart disease, perhaps its only important omission, it was otherwise so comprehensive as to examine the role of smoking in accidents, finding that 18 percent of all deaths from fires in the United States occurred because of the negligent handling of a cigarette, cigar, or pipe.

Was there anything good to say about the leaf? Sort of. The report devoted a page and a half of its 387 to "possible benefits" of tobacco, and declared that people often utilized it as part of "a psychogenic search for contentment." Would smoking help them find it? Possibly, for a few moments here and there, but tobacco's effects in this regard could not be quantified and were, to put it mildly, overwhelmed by charges on the other 385-and-a-half pages.

About pipes and cigars, although unable to say anything good, the report did manage to state that the news was less bad. Whereas cigarette smokers had a mortality rate 68 percent higher than that of non-smokers, the figure for cigar users was 22 percent and for aficionados of the pipe but 12 percent.

Approaching its end, the report addressed the issue of tobacco's continuing popularity in the face of so much opposition from so many reliable sources. It stated, in summary, that "the habitual use of tobacco is related primarily to psychological and social drives, reinforced and perpetuated by the pharmacological action of nicotine on the central nervous system." In other words, tobacco is addictive, and once people get hooked on it, even their knowledge that the substance is harmful may not be enough to unhook them.

All of these qualities, *Smoking and Health* asserted, combined to make cigarettes "a health hazard of sufficient importance in the United States to warrant appropriate remedial action." But the report did not define "appropriate remedial action," nor did it set a timetable for it or even demand that it be undertaken. This failure to be more assertive, and only this in the entire document, would prove to be a source of chagrin to the anti-tobacco forces.

At eleven o'clock that Saturday morning in January 1964, Dr. Luther Terry walked out onto the stage of the State Department auditorium and, stationing himself behind a podium, began to adjust his microphone. Around him on the walls were nine "No Smoking" signs that had been put up for the occasion; it would be almost impossible to take a picture of Terry without also photographing the admonition. The surgeon general tapped the mike, cleared his throat, and stared out at the journalists. He said he would be happy to answer questions. Cameras rolled and clicked; new pages were flipped over in the notepads as hands shot up and voices competed to be heard.

The doors of the auditorium had been locked behind the reporters as they entered so they would not be interrupted while they read. Outside, some men in dark suits and darker visages paced the lobby and corridors of the State Department. They gulped down so many cigarettes that the official U.S. government ashtrays overflowed with butts, and the waste cans with crinkled packs. The men were tobacco industry officials and spokesmen. They knew what was going on in the auditorium and were showing their stress in the time-honored way. "Some puffed determinedly," wrote Marjorie Hunter, who was there, keeping a vigil, "as if trying to convince themselves that they were not a bit worried about the report's conclusion that cigarette smoking enhances the death rate." A few of them listened at the doors of the auditorium; others kept their distance, not wanting to know, although they already did.

Behind the doors, a reporter asked the surgeon general what advice he would give to people in light of the report. "Speaking as a doctor," Terry responded after a moment's thought, "I would tell a youngster not to begin smoking. I would tell an adult smoker to stop smoking, and if he persisted in smoking, I would advise him that he was running a health risk in doing so."

Meanwhile, in the lobby, the tobacco men kept pacing. How much longer was this thing going to last? How much more was going to be said about America's grand old habit?

In the auditorium the questions kept coming, and the author of the present volume wonders about the men and women who asked them. How much, if anything, did they know of tobacco's past? Did they realize that the surgeon general's report was a repudiation of some of history's most dashing figures? Did they know it was vindication for some of its most ridiculed?

Sir Walter Raleigh sold tobacco to England, but James I had a better sense of where it would lead.

John Rolfe developed tobacco in America, but William the Testy was right to bemoan its presence.

George Washington grew tobacco and traded it and demanded it for the soldiers he commanded, but Benjamin Rush, who never rose above the status of historical footnote, was the father of his country's eventual understanding of the leaf.

Old Put craved tobacco and brought cigars back with him from Cuba and spread the word about them to all his friends, but Horace Greeley's comment about a fire at one end and a fool at the other was another word worth spreading.

Andrew Jackson chewed tobacco and spit out the juices and helped to lead an entire nation into similar habits, but Dickens and Trollope and de Tocqueville and other foreign visitors, never having seen such behavior before in their lives, were justified in their disgust.

Wall Street wheeler-dealers and Main Street flappers and other glamorous figures of the Roaring '20s brandished their cigarettes and posed for glossy photos and stood so well for the frenetic spirit of the age, but Lucy Page Gaston spoke the truth.

Movie stars puffed their smokes on screen, passing them back and forth to one another, almost heroic in their gestures, but to modern eyes there is something comical in their gravity. When Paul Henreid slips not one but two coffin nails between his lips, lights them, and passes the spare to Bette Davis, it is supposed to be a dramatic high point of *Now, Voyager*. Today, it is pure schmaltz, a nadir of unintended comedy. It was, rather, Sir Ernest L. Kennaway and the other colorless scientists in the labs, men whose faces would never appear on the silver screen or their names on a marquee, who would be the heroes to posterity, whether posterity knew their names and acknowledged their diligence or not.

In retrospect, then, the report of the surgeon general, *Smoking and Health*, released to the American public on January 11, 1964, may be seen as the ultimate revenge of the nerds.

THE TOBACCO INSTITUTE, the trade council representing America's cigarette companies, had been monitoring the surgeon general's investigations for some time and knew the report would be critical. It had braced itself for the longest list of charges yet, as well as the most damning. But the institute had not expected anything like this. It responded weakly, with what Bob Woodward and Carl Bernstein, in the course of their Watergate inquiries several years later, would call a "non-denial denial"—that is, a rebuttal that does not rebut so much as try to change the subject.

George V. Allen, president of the Institute, even seemed to take the same side as *Smoking and Health* when he said, "I endorse wholeheartedly and without any reservation Surgeon General Terry's call at his press conference today, not for less, but for more research—by the Public Health Service, the American Medical Association, and other public and private groups of scientists who are seeking the scientific facts we so urgently need."

What Allen was really saying, of course, was that those facts had not been provided by the surgeon general's report, which was in his view nothing more than what the media had initially been expecting: old allegations, a new package. The 387 pages, Allen went on, "simply evaluated and reprocessed other studies" and, as a result, were "not really the last word on smoking and health."

That much was true. But most of the words to follow, and there would be torrents of them now as opposed to the trickle of years past, would be even worse.

ON JANUARY 11, 1964, the sixteen hospitals run by the U.S. Public Health Service, an agency supervised by the Office of the Surgeon General, were still distributing cigarettes to their patients—a little something to cheer them up, get their minds off their lung disorders and coughing jags and irregular heartbeats at the same time that their conditions grew steadily worse. The fifty or so hospitals that the Public Health Service helped to operate for native Americans were also handing out smokes.

By January 12, it had occurred to at least a few associates of the surgeon general that this practice, which had always been a bad idea, was now a worse one. Almost immediately it was halted. Not only did the government stop dispensing cigarettes to those in its care; it forbade them to purchase and smoke their own while hospitalized. Furthermore, "Staff members at these hospitals were ordered to conduct educational programs to discourage smoking" and, of course, to refrain from lighting up anywhere on the premises themselves.

ON JANUARY 11, 1964, there was no such thing as a warning label on a pack of cigarettes. In fact, even by the end of the year no remedial action of any sort, appropriate or otherwise, had been taken, and the surgeon general's people were beginning to fear that perhaps the weed would withstand the assaults of science yet again. They were frustrated, angry; a few of them even wrote a song about their feelings and, with cups of punch in hand and voices booming rather than melodious, they sang it at their agency's annual Christmas party, more than eleven months after the report had been issued:

On top of Old Smoking,
A year has gone by,
But the smoke we're deploring
Still gets in your eye.

Then, the following summer, Congress passed the Cigarette Labeling and Advertising Act, its first official response to *Smoking and Health.* President Johnson, who as a younger man had smoked "about three packs a day, often lighting one cigarette while another was still burning in the ash tray," whose "fingers were stained yellow with [nicotine]," who once asked a doctor whether he would be able to keep smoking if he had suffered a heart attack and, when told he would not, responded, "I'd rather have my pecker cut off"—President Lyndon Baines Johnson signed the measure with mixed emotions. It was a good thing for smokers, he believed, especially younger ones, but would they realize it? As for non-smokers, they would support him, but tobacco farmers were his constituents, too. What would *they* do, especially at the polls?

The act took effect on January 1, 1966, almost nine years after an attempt to pass a similar measure had failed. This one required every pack of cigarettes manufactured in the United States to carry the message that the enclosed goods "may be hazardous to your health." No other product, whether in America or anywhere else in the world, had ever warned consumers against its use.

Yet the wording pleased no one. The tobacco industry did not want a warning at all, although, expecting worse, it was less annoyed than it had expected to be. As for the surgeon general and his allies, they had hoped for stronger language and saw the warning as an example of political compromise at its worst. So did the *Atlantic Monthly*, editorializing that tobacco manufacturers had, after all this time, "found the best filter yet—*Congress.*"

And, in fact, very few people seemed warned by the warning. Cigarette sales increased by 16 billion in 1965 and rose almost 8 billion more the following year.

But in time, the stronger language came, and so did the results for which the surgeon general had hoped. After a few years, the "may be" was changed to "is" and the "hazardous" to the slightly more menacing "dangerous." Then the warnings went into newspaper and magazine ads for cigarettes as well as onto the packs. And starting in 1985, they got specific, with a variety of different advisories stating that tobacco caused lung cancer, emphysema, and heart disease; that it might result in fetal injury, premature birth, and low birth weight if used by pregnant women; that, once ignited, a cigarette produced the deadly poison carbon monoxide; and that even if a particular brand of cigarettes happened to be free of additives, it was still not a safe product.

ON JANUARY 11, 1964, there was no such thing as an anti-smoking message on American television. But the surgeon general wanted one. His report had urged the formation of something called the National Clearinghouse for Smoking and Health, the mission of which was to explain the dangers of smoking to the general public through both print and broadcast media.

Congress, however, had clipped its wings, approving the agency but then allocating 2 million dollars a year to run it, a sum that worked out to less than 1 percent of what the cigarette firms were spending on their own media messages of a very different sort. To the surgeon general's office, this was not just another political compromise, but worse: It was a total capitulation to the leaf lobby. The National Clearinghouse was a failure; legislators from the tobacco states made sure it stayed that way.

Before long, though, private sources of money were found, with several philanthropic and charitable organizations kicking in and some private individuals making up the rest. As a result, the case against tobacco finally got some air time, in thirty- and sixty-second versions. Some of them featured announcers reading straightforward copy over pictures of unsavory looking smokers. In others, statistics were recited, and predictions were made for those who ignored them. And in still others, celebrities like Tony Curtis and Tony Randall pleaded with people not only to give up cigarettes but to recruit for the cause by wearing buttons, as they did, that said, "I Quit."

The most dramatic entreaty, though, was filmed by William Talman, a character actor known to millions as Hamilton Burger, the district attorney bested every week by defense lawyer Perry Mason in the TV series of the same name. Talman wrote the message himself, and the complete script, transcribed by this author from a videotape recording, follows. As you read it, you must picture the man with a face far more gaunt and lined, eyes far more sunken, and complexion far more sallow than they ever appeared in Perry Mason's make-believe courtroom. You must try to hear the voice of a man who is not just reading lines, but speaking his heart. This, rather than the contrived and exploitative police raids and the over-hyped talent shows and the desert-island survival programs of the late twentieth and early twenty-first centuries, is the real reality television.

60 SECOND WILLIAM TALMAN
TV SPOT
16mm Color SOF

(scenes of Talman family in various phases of activity around the house)	This is the house we live in. That's Billy; he's pretty handy to have around. Steve, home from college; Barbie, looking after her brother Timmy; Debbie, who'll soon graduate from high school; Susan, who has captured all our hearts; and my wife Peggy, who looks after all of us.
(closeup of photo of Talman and Raymond Burr, who played Perry Mason)	And that's me—Bill Talman, with a friend of mine you might recognize. He used to beat my brains out on TV every week for about 10 years.
(medium close up of Talman in den)	You know, I didn't really mind losing those courtroom battles. But I'm in a battle right now I don't want to lose at all because if I lose it, it means losing my wife and those kids you just met.
(closeup of Talman)	I've got lung cancer. So take some advice about smoking and losing from someone who's been doing both for years. If you haven't smoked—don't start. If you do smoke—quit. Don't be a loser.

Shortly after this announcement began to run on television stations all over the United States, William Talman died. He was sixty-one years old. As per his request, the message continued to be broadcast, the dead man doing his best to reason with the living about the subject that had so consumed, and caused, his final days. There was at least one newspaper report of a woman giving up smoking because, seeing Talman on the tube one night and knowing he had passed away, she thought she had seen a ghost, and she was not about to ignore so imposing, if spectral, a presence.

ON JANUARY 11, 1964, there was no such thing as a ban on television ads for cigarettes. In fact, as a result of *Smoking and Health*, the tobacco industry, desperate to blunt the report's effects to whatever extent it could, became the largest of all TV advertisers, eventually accounting for 8 percent of commercial time on the three major networks. It was the same thing the companies had been doing for decades: reacting to serious charges with frivolous claims, some of them for improved filters, which were not; others for new, low-tar smokes, which were not low enough to make a difference; some for mentholated brands, which attempted, against all logic, to make hot smoke taste cool.

But this time Congress ignored the tobacco lobby and acted in what it believed was the public good. It ruled that, as of midnight, January 1, 1971, there would be no more tobacco ads on the American tube. Cigarette makers were stunned. To some it was a far more serious blow than either the surgeon general's report or the warnings on the packs. Many smokers either knew of or suspected the information in the former and ignored the latter. But without any more television exposure, the tobacco companies feared it might be cigarettes themselves that people began to ignore.

So they did everything they could, legal and otherwise, to influence the vote. They threatened and promised, cajoled and bribed. They hired extra staff, worked extra hours, turned out extra letters and pamphlets and petitions. They were as frantic as Lucy Page Gaston looking down the barrel of World War I, and for a time it seemed as if they might actually succeed. Debates over the ad ban in both the House and Senate were long and bitter, and some people were predicting they could go either way. Eventually, though, they went the surgeon general's. The tobacco firms had lost another one.

The date, at least, was a victory for them, as lawmakers decided to give the industry one last fling, a chance to pitch their wares one last time before the huge audiences on New Year's Eve and the following day during the college football bowl games. The companies responded by blowing the doors off their vaults and storming them with fleets of wheelbarrows. Philip Morris alone spent more than a million dollars for ads on December 31 between eleven-thirty and midnight, bingeing on self-promotion as so many of the company's intended customers were bingeing on alcoholic beverages to welcome the new year.

Then the next day, as Stanford beat Ohio State in the Rose Bowl and Nebraska nipped LSU in the Orange and Notre Dame topped Texas in the Cotton and Tennessee got past the Air Force Academy in the Sugar, American tobacco companies paid out more money to pitch their products than they had ever invested in a single day before. Perhaps in a single week. There was as much cigarette smoking on the air as there was running, passing, and kicking. Never again would the companies have a chance to celebrate themselves in such a manner; they wanted the impression to endure.

But as things worked out, it was not the end of cigarette advertising on television—not really. It was, rather, a change of format. The leaf companies had figured out ways to work around the ban almost as soon as it was announced. They were like accountants finding loopholes whenever the tax code is revised. One of them was to start subsidizing athletic events or the entrants in those events. A trade publication called *Tobacco Reporter* explained:

> For instance, at a stock car race one or more of the participating vehicles will have a cigarette promotion blurb on its side. In following the race the television cameras will pass fence advertisements while the crews and winners will be smoking with the cigarette prominently displayed. Such will hold true for any number of sporting events that appear on television. It could even go so far as to have the football booster section displaying block cards that promote a particular brand.

It did not go that far, and the sponsorship of sporting events was not as effective as previous, more direct advertising. But it was cheaper, even with the cigarette companies putting up huge pots of prize money to attract the top athletes in their sports so that the television networks would in turn be attracted. And it was legal, completely on the up-and-up, and no greater advantage than this now existed.

Tobacco companies also made sure that their brand names were integral parts of the names of the events, so that whenever the TV announcer reminded viewers that they were watching, say, the Virginia Slims Tennis Tournament, he was, in effect, doing a little commercial, driving home the point. And when the tournament was over, a Virginia Slims executive stood before a Virginia Slims banner to hand the winner a Virginia Slims check and stood by as the winner thanked all the nice folks at Virginia Slims for so generous an amount.

In 1970, the American tobacco industry spent a total of $500,000 to organize, run, and publicize athletic contests. They were a sidelight in those days, an afterthought. Five years later, with the advertising ban having become a fact of life, the amount had increased seventeen-fold—to $8.5 million. They might not be ideal, but sporting events were among the best of the venues that remained for selling cigarettes in the United States.

And movies were not bad, either. Several sources believe that the 1980s were the golden age of product placement for the cigarette industry in feature films, an impression confirmed by a cursory viewing of the decade's most expensive and widely publicized releases. Rocky Balboa might not have been a smoker, and Rambo might also have prized his physical condition above all other concerns, but in 1983, Sylvester Stallone, who played both characters, proposed a deal to the representative of one of America's leading cigarette makers. "I guarantee that I will use Brown & Williamson tobacco products in no less than five feature films," Stallone wrote. "It is my understanding that Brown & Williamson will pay a fee of $500,000.00. Hoping to hear from you soon."

He heard. The deal was made, as were a number of others. According to a British publication called *Tobacco Control*, and as reported in the *New York Times* a couple of decades later, "At least one company went so far as to provide free cigarettes to actors and directors who might therefore be more inclined to light up when the cameras rolled." The company was R. J. Reynolds. The actors and directors, among others, were Jerry Lewis, John Cassavetes, Liv Ullmann, and Shelley Winters.

ON JANUARY 11, 1964, there were very few non-smoking sections in the country, in office buildings or airplanes or trains or buses or public arenas. In fact, Americans had a history of creating sections *for* smoking. Decades earlier, when women who lit up were scorned in some quarters and legislated against in others, a few places went out of their way to be hospitable:

> Dozens of theaters in New York and elsewhere opened special smoking rooms for women; the amenities included complimentary cigarettes for patrons who had forgotten to bring their own. At the Woods Theatre in Chicago, women could smoke in luxury in a room appointed with $10,000 worth of marble, Persian rugs, and leather furniture.

But since then, much had been learned about what was first called involuntary smoking and is now known either as secondhand smoke or, more officially, environmental tobacco smoke, or ETS. The Environmental Protection Agency believes that ETS causes 3,000 fatal cases of lung cancer a year in non-smoking adults, as well as an estimated 35,000 deaths among non-smokers from heart disease. ETS is also responsible, says the EPA, for bronchitis and pneumonia in infants and for more severe asthma attacks in children who already have the ailment. It even seems to cause asthma in children not previously afflicted—by one estimate, as many as 26,000 cases annually.

And so in 1973 smokers and non-smokers were separated for the first time on commercial airliners. The same year, the Interstate Commerce Commission moved smokers to the back of the bus, literally.

In 1977, cigars and pipes went into segregated quarters on airplanes, as well. Not until 1990 would a total ban on smoking take effect.

The first statewide ban on smoking in public buildings was passed by the Arizona legislature in 1973. Two years later, Minnesota went further; its Clean Indoor Air Act intended to protect "the public health and comfort and the environment by prohibiting smoking in public places and at public meetings, except in designated smoking areas." In 1978, the federal government outlawed tobacco in all of the buildings in which it did its business, and since then most private workplaces have done the same.

Today, restaurants and theaters also prohibit or severely isolate the weed, as do office complexes and stores, libraries and museums, health clubs and country clubs, waiting rooms and meeting facilities, auditoriums and stadiums, and various other places where men and women and children assemble, both indoors and out. As the activity of a minority, smoking is allowed in but a minority of locations in the early years of the new millennium.

ON JANUARY 11, 1964, there was no such thing as a lawsuit against a tobacco company for product liability, much less a successful one. Now, after more hours of litigation and hearings than anyone can count, and more pages of documentation than anyone can decipher, and more fees paid to lawyers than even the most voracious of them can spend on second and third homes and SUVs for use on lacquered suburban highways, the tobacco industry has been defeated several times in several different courtrooms.

In July 1999, a Florida jury found against the industry in the first class-action lawsuit to reach a verdict, ruling that cigarette makers deliberately hid the dangers and addictiveness of their product.

In Portland, Oregon, two and a half years later, a jury even found in favor of a woman whose death was said to have been caused by *low-tar* smokes.

And in perhaps the most notable of all cases so far, the industry agreed to pay 246 billion dollars in damages to the attorneys general of forty-six American states to resolve claims for health costs related to smoking. But the total might well go higher. The federal government has jumped into the pool, as well, announcing a suit of its own that seeks to recover billions more in smoking-related payments made by Medicare and other federal insurance programs. Further, the government wants tobacco companies to forfeit close to 300 billion dollars in profits. As the *New York Times* reported in March 2003,

> The Justice Department asserts in more than 1,400 pages of court documents that the major cigarette companies are running what amounts to a criminal enterprise by manipulating nicotine levels, lying to their customers about the dangers of tobacco and directing their multibillion-dollar advertising campaigns at children.

The *Los Angeles Times* summed up: "During a period of 35 years starting in 1960, cigarette makers won virtually every trial by shifting attention from themselves to the foolishness of smokers, who persisted in their habits despite health warnings. Overall, the cigarette companies have won more than three-fourths of all the cases that have gone to verdict. But the industry's fortunes began to change in the mid-1990s after reams of secret internal documents came to light showing that the companies long knew of the health hazards and addictiveness of their products."

So much, in fact, has the legal tide turned against tobacco companies that, according to anti-tobacco crusader John Banzhaf III in an e-mail to this author, as of August 2003, "16 states now accept the principle that a smoker who deliberately subjects a child to tobacco smoke pollution can lose custody of the child."

As part of the agreement with the states' attorneys general, the cigarette industry agreed to cut back on its sponsorship of sports and entertainment events. It has done so but still provides the backing for a number of other activities, such as music festivals, trade shows and conventions, and dances that are held in adults-only clubs.

Also as part of the agreement, the states were to have committed large chunks of the settlement money for the funding of anti-smoking programs within their borders. As of mid-1999, only six states had complied in meaningful fashion. The rest, it seems, have been using substantial percentages of the tobacco fines to balance state budgets or pay for programs of some other kind.

ON JANUARY 11, 1964, there were billboards advertising cigarettes from one end of the country to the other. They sat alongside superhighways and two-lane blacktops, in the city and in the sticks, and they showed rugged men and elegant women lighting up and drawing in and blowing out, obviously delighted to be smoking, their delight meant to urge those who looked at them to reach for a smoke themselves, perhaps even as they were driving by. The men and women on the billboards were shown sitting on beaches and hiking up mountain trails and sitting atop horses and dining in the fanciest restaurants. The social cachet of cigarette use was obvious to a motorist at any speed, certainly to a pedestrian. And the size of the billboards was a reinforcement of the image. These were the big-screen plasma TVs of the American roadway.

But another stipulation of tobacco's pact with the forty-six states: The industry had to get rid of its outdoor advertising, all of it, and promptly. By midnight, April 22, 1999, every billboard in the United States hawking cigarettes had come down, leaving Joe Camel and his crowd with almost nowhere to hide but the pages of newspapers and magazines, and finding not nearly as many of those available to them as there used to be.

In fact, in May 1999, the *New York Times* became the first national paper, joining more than a dozen local publications, to reject all advertising for tobacco products of all kinds. The cigarette companies cried foul. A "pathetic" decision, said one spokesman. Whatever happened to "freedom of commercial expression?"

Arthur O. Sulzberger Jr., publisher of the *New York Times*, responded by pointing out that, as far as he knew, there was no such thing and never had been. The First Amendment, he said, "gives the press the right to publish what it chooses to. It doesn't force the press to publish something, whether that's a news story or an advertisement."

ON JANUARY 11, 1964, a little more than half of American men and a third of American women smoked cigarettes. By the fall of 2003, the figures had dropped to 25 percent and 21 percent, respectively. As for per capita consumption, it has also decreased markedly, from a peak of 4,345 smokes in 1963 to an average of 2,423 in 1997.

Even soldiers were lighting up less than they used to. In 1980, the percentage of men in uniform who identified themselves as casual smokers was 51. By 1998, it had sunk to 30. The percentage of heavy smokers also declined sharply. Military actions in Vietnam, the Persian Gulf, Afghanistan, and Iraq came and went (and, in the latter case, has been resumed) without any generals demanding smokes for their men or warning of the consequences if they were denied.

But boys and girls under the age of eighteen were smoking more than before, possibly more than ever. Between 1991 and 1997 there was a 32 percent increase in cigarette use among American high-school students. In 2003, the American Cancer Society reported that three hundred young people were picking up the cigarette habit *every day*. They were smoking on school grounds even when it was not allowed; they were smoking on the sidewalks of town and on the paved plains of shopping malls even when adults looked askance; they were smoking at private parties, where first- and secondhand smoke can mingle indiscriminately, working all manner of insidious effects on them. As researcher M. A. H. Russell has learned,

> The modern cigarette is a highly effective device for getting nicotine into the brain. The smoke is mild enough to be inhaled deeply into the alveoli of the lungs from where it is rapidly absorbed. It takes about 7 seconds for the nicotine absorbed through the lungs to reach the brain compared to the 14 seconds it takes for blood from arm to brain after an intravenous injection.

Perhaps teenagers are too young to fathom the limits of mortality, and the painful steps that can bring it to a premature close. Perhaps the pressures of growing up today are greater than they used to be and make teenagers seek release in the same way that soldiers used to seek release from the strains of battle. Perhaps they just need something to rebel against, and good sense, as often seems to be the case, is the current target.

As for cigars, they are enjoying a revival of sorts. In 1993, they reversed a twenty-year trend and began to gain in popularity, and the bigger and more expensive the cigar, the greater the increase in sales. The premium, handmade, imported variety, for example, showed a 250 percent jump between 1993 and 1998, in large part because the bull market on Wall Street led to a renewed period of conspicuous consumption in the United States, of which cigars, like designer drinking water and multifunction wristwatches with condominium-size price tags, are a perfect symbol.

In addition, there has been a backlash against what some people call the "nanny culture," which is to say, what is thought to be an excessive concern on the part of institutions for the welfare of individuals. Which is to say, meddling. Adults, too, can feel the urge to rebel against good sense, especially if enough people are preaching it with enough militancy—and cigars—which most people smoke infrequently and seldom inhale, are a safer means of lashing out than cigarettes. In fact, in his book *The Pleasure Police*, cigar fancier David Shaw rejects the notion that good sense is being disregarded at all:

> Even if there is some mild health risk, I am reasonably certain that for someone like me, with a family history of heart disease and a tendency to rush around trying to do three things at once, that minimal risk is more than offset by the completely stress-free, anxiety-free hour that it takes to smoke a cigar while doing absolutely nothing else. It's a lot better than psychotherapy—and about $100 an hour cheaper.

ON JANUARY 11, 1964, there was no such thing as an *anti*-smoking industry in the United States. There were a few products here and there that swore they could help people quit; there were a few treatment centers and the occasional book explaining how thought control and self-hypnosis could wean even the most avid smoker from his dependence on the leaf. But there was nothing extensive, nothing well known, nothing that was widely acknowledged to be reliable.

Today, the companies offering a cure for tobacco seem to be thriving as much as their adversaries thrived in ages past. They sell chewing gum and skin patches, nasal sprays and inhalers and beverages, most of which provide a substitute source of nicotine without the other harmful elements of cigarette smoke and are designed, through decreased usage, to reduce a person's dependence on nicotine as time goes on. In addition, a person can buy anti-smoking pills at the pharmacy, can sign up for hypnosis through groups such as Positive Changes, and can partake of anti-smoking support groups based to some extent on Alcoholics Anonymous. Some people have taken to cigarettes made of lettuce; others believe in acupuncture or vitamin therapy or little machines that punch tiny holes in cigarettes in an attempt to dilute the poison.

Those who succeed in giving up the weed can be well rewarded. Studies have shown that a smoker who has been separated from his smoke for a mere twenty-four hours has already lessened his odds of a heart attack. After a month or so, his lungs are functioning as much as 30 percent more efficiently than they were before. By the time a year has passed, his chances of a heart attack are cut in half. At the five-year mark, his chances of a stroke drop significantly. Ten years: His risk of dying from lung cancer is about half as great as that of a smoker. Fifteen years: His risk of coronary disease is the same as that of a man or woman who has never put a cigarette between his lips in his life.

ON JANUARY 11, 1964, there was no such thing as a tobacco company calling itself something other than a tobacco company—and, in truth, *being* something other than solely a tobacco company. But so much else has changed in the four decades since the surgeon general's report; it is no surprise that this did, too.

The American Tobacco Company, which had for many years boasted in its advertising that "Tobacco Is Our Middle Name," has long since dropped the word entirely. On July 1, 1969, it became officially known as American Brands, Incorporated.

Not long afterward, the Reynolds Tobacco Company also changed its named. It is now R. J. Reynolds & Company.

In neither case, nor in the case of other firms doing the same thing, was the company adopting a pseudonym and thereby trying to deceive—not primarily, at least. Rather, the new names were labels that reflected a new corporate reality. Cigarette makers had been forced by adverse publicity, continuing public pressure, and disappointing sales to come up with new product lines, different product lines, to stop relying exclusively on the weed. They could see that smoking was on the way down, feared it was on the way out, and assumed they would have to sell other merchandise besides cigarettes if they hoped to stay solvent. Diversification. It is a common business strategy today; back then, at least for the tobacco industry, it seemed innovative, perhaps even risky.

In 1970, Liggett & Myers joined the name changers. It took out a full-page ad in several newspapers to announce that it was no longer just a tobacco company. Henceforth, it was a pet food, liquor, wine, cereal, popcorn, watch band, *and* tobacco company. Liggett & Myers was playing the odds. Surely, thought the folks in the carpeted corner offices at corporate headquarters on the top floors of gleaming skyscrapers, no one could turn up any dirt on watch bands.

The strategy worked, and not just for Liggett & Myers. Even though the doctors and politicians and journalists were against the tobacco companies—which is to say that they had lost the fact finders, lawmakers, and opinion shapers—they managed to keep a fair number of their old customers. And by diversifying intelligently, they gained new customers for their new products. The partial-tobacco entity Philip Morris, for example, is today one of the thirty firms that provide the Dow Jones Industrial Average. Another partial, American Brands, is not only the second-largest tobacco company in North America but one of the top ten beverage conglomerates.

And so the companies that make cigarettes make billions of dollars a year in profits. Their stock prices rise; their shareholders reap dividends that are sometimes unexpectedly high; their top executives get raises and bonuses that are, on occasion, larger than their original yearly salaries, with twelve months of employment often providing them with more money than most human beings make in a lifetime. But this is not the true measure of tobacco's effect on the American economy. In fact, it is an illusion, as harmful to one's perceptions as a cigarette is to one's physical well-being.

According to the American Cancer Society, 400,000 people die every year in the United States from tobacco-related causes. It is also believed that tobacco is responsible for as many as 18,000 miscarriages a year. Health-care expenses directly attributable to the weed in one form or another total more than 50 billion dollars annually. Lost hours on the job and similar, smoking-induced inefficiencies cost another 50 billion dollars, if not in excess of that. The leaf, as these figures have it, siphons off 100 billion dollars' worth of U.S. productivity every year.

Granted, the American Cancer Society might not be the most objective source of information about the tobacco industry. And granted, the numbers it provides are so large and rounded-off as to seem inauthentic—or, at least, terribly imprecise. Nonetheless, the general conclusion seems unassailable: Tobacco costs Americans a lot more money each year than it brings in. It is a deficit business in a growth nation, a source of profit only to itself, and this only because the individual firms sell so much more than just the products that got them started.

ON JANUARY 11, 1964, there was no such thing as a tobacco company that admitted the connection between smoking and cancer. Now there is. In the winter of 2004, the Philip Morris website announced that the company "agrees with the overwhelming medical and scientific consensus that cigarette smoking causes lung cancer, heart disease, emphysema and other serious diseases in smokers." It agrees that "there is no safe cigarette," including the low-tar brands. And it says that "to reduce the health effects of smoking, the best thing to do is quit; public health authorities do not endorse either smoking fewer cigarettes or switching to lower-yield brands as a satisfactory way of reducing risk." Other tobacco companies have made similar admissions. It is, to some, the equivalent of the devil's admitting a downside to sin.

But Philip Morris had not suddenly turned compassionate. Nor had it suddenly become astute in interpreting medical data. Its admissions were, or were intended to be, good business. By acknowledging that their products were killers they hoped to avoid lawsuits on behalf of people who had died claiming the dangers of smoking were unknown to them, despite the pervasiveness of warnings.

They also hoped to avoid prosecution by the federal government under the provisions of the Racketeer Influenced and Corrupt Organizations (RICO) Act. The government has stated that "the companies have deceived the public on matters of smoking for more than 50 years." In response, a Phillip Morris attorney, referring in part to the firm's website, says, "The core theme of our defense is that during the last several years there has been such profound and fundamental change in how tobacco companies communicate with the American public about the risk of smoking that there is no likelihood of future RICO violations."

Truth in advertising from an American cigarette maker, however self-interested the motive, at long, long last.

ON JANUARY 11, 1964, Wayne McLaren was still alive. In fact, he was only twenty-three years old, robustly healthy and ruggedly handsome. He had not yet appeared in such movies as *Paint Your Wagon* and *Butch Cassidy and the Sundance Kid*, had not worked as a stunt double in other movies for Burt Reynolds, had not been a model for United Airlines and Ford trucks. And he was more than a decade away from becoming the most familiar of all the faces who portrayed the Marlboro man in print ads and on billboards. It was one of the most popular images in the history of advertising, an image that, according to B. A. Lohof in the *Journal of Popular Culture*, "represents escape, not from the responsibilities of civilization, but from its frustrations. Modern man ... jealously watches the Marlboro Man facing down challenging but intelligible tasks."

But there was a task too challenging for Wayne McLaren to face down. He could not save himself from the ravages of lung cancer, the result of a pack and a half of Kools—not Marlboros—a day for three and a half decades. He died in the summer of 1992, at the age of fifty-one.

By that time, he did not have much of a voice left; he had had surgery on his larynx and could not speak normally. But for several years he had been speaking with abnormal passion about his life as a smoker, making public appearances to dissuade children from the habit, addressing the Massachusetts legislature about the wisdom of anti-smoking laws, and appearing in a British Broadcasting Company documentary called *The Tobacco Wars*. "I've spent the last month of my life in an incubator," McLaren whispered, near the end, "and I'm telling you, it's just not worth it. I'm dying proof that smoking will kill you."

ON THE DAY that Surgeon General Terry released his report, Edward L. Bernays was still proud of himself—or, at least, not embarrassed enough to make a public act of contrition. He had won fame, riches, and honorary degrees galore. A behind-the-scenes kind of guy, he had become a denizen of center stage and was as comfortable with the limelight as he had once been with the planning session behind closed doors. He was even admired, at least grudgingly, by some of those who found the fruits of his labors poisonous. He was the master of an entire field of human endeavor, and to one extent or another, all those who followed him into the field would be his pupils.

But the surgeon general's findings were a blow to Bernays, something almost personal. They solidified doubts that had already begun to plague him and that, in the months following, became a source of anguish. Bernays had to do something, had to come up with some kind of idea. He racked that fertile brain of his for a means of making amends. Late in 1964, he thought he had it.

Edward L. Bernays declared that he now knew how to turn smoking, the activity he had so cleverly and indefatigably promoted, into "an anti-social action which no self-respecting person carries on in the presence of others." He did not get into specifics, only saying that he would be "enlisting moviemakers and radio and television personalities, advertising executives, clergy and doctors, in a bid 'to outlaw and eliminate cigarette smoking.'"

It did not happen. It was too much. Even the father of spin, the designer of green fashions, and the man who had once sworn to the wisdom of substituting Luckys for sweets, could not bring about a transformation of that magnitude. Almost immediately he scaled back, joining with others to work for more realistic goals, such as a continuing decline in sales of the weed, and he helped to bring about the ban on cigarette advertising on TV and radio. The latter was a great source of satisfaction to him, he said. It "got rid of a sense of guilt" that had been with him for too long, replacing it with peace of mind, a feeling of atonement to which, in the final analysis, Edward L. Bernays was not so easily entitled.

EPILOGUE

"The Ten O'Clock People"

IT IS IN MOST WAYS a typical Stephen King short story. Among the characters are a man, whose face "had been covered with lumps that bulged and quivered like tumors possessed of their own terrible semi-sentient life," and a woman, identically deformed, one of whose lumps "was leaking a thick pinkish substance that looked like bloodstained shaving cream." They are representatives, these two, of a whole clan of subhumans, creatures whose heads are furry and bat-like, "not round, but as misshapen as a baseball that has taken a whole summer's worth of bashing." To make matters even more ghastly, the heads are "*in motion*, different parts moving in different directions, like the bands of exotic gases surrounding some planetary giant." Not exactly the kinds of people with whom you want to share a sauna at the local health club.

But it is not these monsters who make the story relevant to the present volume; it is the tale's title figures. They are "The Ten O'Clock People," and it is they, and only they, who can see the bat-faced subhumans for the fiends they really are. To other mortals, the bat faces appear normal human beings, well-dressed men and women of the business world, occupants of executive suites whose heads are round and stationary, un-lumpy and non-oozing.

The Ten O'Clock People are smokers, so called because every day at ten in the morning, and again at three in the afternoon, they emerge from their office buildings as voluntary exiles, standing in snow or rain, heat or gloom, stiff breeze or air congealed with humidity, nothing staying them from a swift indulgence in their sorely discredited habit. They light up, suck in, blow out, savoring their cigarettes like the pernicious pleasures we now know them to be.

Brandon Pearson belongs to a group of such individuals who work for the First Mercantile Bank and can be found at the appointed hours

> in the broad plaza in front of the bank ... some reading, some chatting, some looking out at the passing rivers of foot-traffic on the sidewalks of Commercial Street ... all of them doing the thing that made them Ten O'Clock People, the thing Pearson had come downstairs and outside to do himself.

But it is not just the bank that employs Ten O'Clock People. As Pearson rides across town in a cab with his friend Duke, it occurs to him "that easily ninety percent of the posh midtown highrises they were passing were now no-smoking zones," and each has its own little band of men and women who need two short shifts a day outside to ease their minds and punish their bodies. "An exotic social group," is how Pearson thinks of his fellow cigarette devotees, "but not one that was apt to last very long. He guessed that by the year 2020, 2050 at the latest, the Ten O'Clock People would have gone the way of the dodo."

For the present, they are men and women bound together not only by their habit but also by the disdain that the rest of society feels for it and, as a consequence, for them. They seek each other out not just to share their nicotine highs but also, perhaps even more important, for companionship and understanding, outsiders joining with other outsiders to form small cliques of temporary insiders. All of them, it seems, have the same regard for their cigarettes that Pearson does. He lights up, "relishing the way the smoke slid into his pipes, even relishing the slight swimming in his head. Of *course* the habit was dangerous, potentially lethal; how could anything that got you off like this not be?"

A few minutes later, Pearson talks to Duke, who is also a member of the group but does not know that Pearson, like an anthropologist, has made a study of its behavior, done some mental cataloguing and classifying. Duke does not even know that his friend has given the outdoor smokers a name:

> So Pearson explained a little about the Ten O'Clock People and their tribal gestures (surly when confronted by NO SMOKING signs, surly shrugs of acquiescence when asked by some accredited authority to Please Put Your Cigarette Out, Sir), their tribal sacraments (gum, hard candies, toothpicks, and, of course, little Binaca push-button spray cans), and their tribal litanies (*I'm quitting for good next year* being the most common).

"Ten O'Clock People," Duke says in a marveling voice, "Man, I love that—I love it that we have a name. And it really is like being part of a tribe."

Which brings us back to the bat faces and the question at the heart of the story: How is it that the Ten O'Clock People are the only ones who can see them? What, if anything, does tobacco have to do with it?

Well, it seems that the stars of the tale are not just tobacco junkies; they are tobacco junkies of a particular kind, all of them having once consumed several packs a day but having now cut back to five or ten cigarettes each twenty-four hours. This, in ways known only to Stephen King, and perhaps not even to him, has altered the body chemistry of the smokers in such a way that they now possess a power they would rather be without. They can see the truth of the bat faces while all others, heavy smokers and non-smokers alike, view them as typical denizens of the nine-to-six world.

The bat faces are not happy about this. They want to punish the Ten O'Clock People for their perceptiveness by devouring them, literally breaking them into pieces and making them into meals. In fact, they are genetically programmed to do just that. Duke gives Pearson some details, sketchy though they are, telling him the creatures "eat something that our brains make, that's what Robbie thinks. Maybe an enzyme, he says, maybe some kind of special electrical wave."

Later, after having mulled this intelligence over for a while, trying to make some sense of it, Pearson asks Duke whether similar things are going on in other countries.

Duke doubts it.

Pearson wonders why.

"Because this is the only country that's gone bonkers about cigarettes ... probably because it's the only one where people believe—and down deep they really do—that if they just eat the right foods, take the right combination of vitamins, think enough of the right thoughts, and wipe their asses with the right kind of toilet-paper, they'll live forever and be sexually active the whole time."

King is more concerned with mayhem than with cultural observation in "The Ten O'Clock People." It is what he does for a living. But there are moments when he paints a sweetly sympathetic portrait of the American smoker at the end of the twentieth century and beginning of the twenty-first, providing an emotional context, however superficial, that gives the mayhem some resonance. King knows something about the men and women and kids who smoke their 2,423 cigarettes a year, and rather than being offended, he understands the attraction. He knows well the angst of quitting and the further angst of failing, perhaps repeatedly, in the attempt. He laments the fact that smokers are outcasts at the same time that he seems to concede the inevitability of their status, perhaps even its desirability for the good of society. The smokers in his story, however ruefully, accept their fate, which plays as a tacit admission of their weakness.

But there are moments in "The Ten O'Clock People" when this reader was cast back in time and found himself recalling the first Europeans to smoke cigarettes. These were the poor people of the cities in the Middle Ages, the ragged individuals who made their own, scavenging through the garbage of the better off, collecting wastes from pipes and cigars and snuff, wrapping them in bits of paper that had also been discarded by others. These were the pariahs who waited until night, when darkness would shield them from the eyes of their social superiors, and then gathered not in stylish places like the First Mercantile Plaza but in out-of-the-way corners, in shadowy recesses, where they puffed away with others of their own kind until dawn.

The Ten O'Clock People of today are the direct descendants of the midnight people of centuries past. They take their tobacco in quarantine and even then know they are being watched, talked about; heads shake at them and eyes narrow, lips curl down in displeasure. The cigarette smoker has come full circle, from rejection to acceptance to rejection.

Unlike the Mayas, the midnight people did not believe smoking would bring salvation. They knew that they gods were occupied elsewhere, and when the smoke disappeared it simply ceased to exist.

It is the same with the Ten O'Clock People. They do not brave the elements and the ostracism of others to transmit their prayers. Some of them, in fact, will even admit the unholiness of what they do, lamenting it at the same time that they yield to it.

Unlike numerous native tribes of the ancient world, the midnight people did not think the weed would cure their diseases, relieve their suffering in any way.

The Ten O'Clock People are also aware that this is so. The innocuous-looking brown plant once thought to be a gift of the gods has, with the passage of centuries, been revealed as a curse, having consigned millions of its victims to discomfort, suffering, and premature death.

Unlike their contemporaries who smoked pipes or cigars or wedged chunks of snuff into their sinuses and then snorted, the midnight people did not dream that cigarettes would either confer prestige or provide a favorable image.

The Ten O'Clock People have learned that what the small smokes bring is exclusion, but although this is at times painful to them, it may also be one of the reasons they continue with their habit: for the satisfaction of flouting the conventions that shun them. If they cannot achieve a higher standing in society because of their addiction to cigarettes, they will at least insist on a distinct identity.

The Ten O'Clock People will endure. Pearson's prediction to the contrary, they will not have gone the way of the dodo by 2020 or 2050. There will always be those so eager for the consolations of the moment that they accept, or perhaps just refuse to admit, a later reckoning. There will always be those who find the cigarette the perfect medium for the statement they want to make, or if that is too grand a way to look at it, for the need they have to fill. There will always be those to whom the smoke itself is mesmerizing in a way that cannot be defined and can barely be described, those who find it the freest of spirits, the embodiment of a whim, a material for dreams, for yearning.

And perhaps this is part of the explanation for tobacco's continued acceptance by some. Perhaps, at some level, it is not the nicotine in a cigarette that traps them, not the tars or the taste or any other combination of the leaf's more tangible enticements. Perhaps it is the fumes, the smoke, so liberated in its paths, so endless in its possibilities. There is nothing quite like it on Earth, never has been, and to produce it from one's own body, through one's own exertions, must be, at least for some people, the most exhilarating of sensations.

It may be that, at least to some extent, this is why they still cling to the weed in the face of so much evidence to the contrary. It may be that this is why they face the censure of friends and strangers alike, not to mention a certain lack of sympathy from the doctors who will one day treat their breakdowns without offering much hope. It may be that their spirits are uneasy, their inner children aged, and their private demons outed—and they cannot help but seek release, however briefly and perilously, in the runaway independence of smoke.

ACKNOWLEDGMENTS

I DID MUCH OF THE research for *The Smoke of the Gods* at the Library Relating to Tobacco, which was assembled by George Arents for the New York Public Library. It is a thorough and enlightening collection of both primary and secondary materials, and this book would not have been possible without them.

For a number of reasons, the research I did elsewhere consists primarily of secondary materials, to which I applied the same rule that journalists are supposed to apply to their reportorial pursuits whenever possible: I used no information that could not be verified by a second and, in many cases, a third source. *The Smoke of the Gods* was written, in other words, in the same manner as its companion volume, *The Spirits of America.* To the best of my knowledge, no errors of fact were ever discovered (certainly none were ever called to my attention) in *Spirits*; I hope the same will be true of these pages.

I am grateful to Professors Ira Berlin and George Calcott of the University of Maryland for discussing with me the relationship between tobacco and slavery.

I am grateful as well to the staff of the Westport, Connecticut, Public Library, who not only found virtually all of the arcane books I requested, but suggested a few of their own. On the library's back shelves is a collection of obscure pamphlets published several decades ago by the Tobacco Institute, which tells much about the crucial, and greatly misunderstood, role of tobacco in colonial times.

My thanks also go to Micah Kleit, Gary Kramer, and Ann-Marie Anderson of Temple University Press, as they do to my ever heartening agent, Tim Seldes.

Most of all, I am thankful that my wife, son, and daughter do not smoke.

NOTES

Introduction: The Ancient World

page 2 "a sense of divinity," Keesecker, ed., p. 40.
3 "portable altar," Reader's Digest Association, p. 142.
3 "most ingenious," Beaumier and Camp, p. 133.
4 "Plants whose properties," Robicsek, p. 27.
5 "They have many odoriferous," quoted in Heimann, p. 9.
5 "It openeth," quoted in Shenkman and Reiger, p. 16.
5 "Unlike herbal remedies," Milton, p. 163.
5 "I am aware," quoted in Robicsek, p. 38.
6 "the most sovereign," quoted in Fairholt, pp. 8–9.
6 "if one but swallow," quoted in Priestley, p. 149.
6 "to resuscitate people," Brongers, p. 29.
7 "Injections of tobacco," quoted in Larson, p. 211.
7 "to drive out," Robicsek, p. 36.
7 "The remedy for snakebites," quoted in ibid., p. 37.
8 "If the illness," ibid., p. 35.
8 "using … leaves," Gately, p. 13.
8 "one could live," Beaumier and Camp, p. 134.
9 "Whenever a pregnant woman," quoted in Brongers, p. 194.
9 "where it was introduced," Gately, p. 13.
10 "three painted and gilded," quoted in Heimann, p. 20.
11 "great physical advantages," quoted in Bain, p. 98.
11 "may have originated," Heimann, p. 21.
11 "Some tribes," Gately, p. 10.
12 "The tobacco must," Heimann, p. 238.
12 "is as much to say," quoted in Heimann, p. 122.
12 "a fine white buckskin dress," quoted in Barth, pp. 31, 33.
12 "filled the pipe" and "Gitche Manito," in Longfellow, pp. 144–45.
13 "suck so long," quoted in Gately, p. 31.
13 "tried a few drags," Morison, *Northern Voyages*, p. 418.
13 "There is nothing more mysterious," quoted in Reader's Digest Association, p. 142.

One: The Old World

page 14 "obscure but ambitious," Wilford, p. 29.
14 "considered a little touched," Morison, *Christopher Columbus*, p. 12.
14 "the shape of a pear," quoted in Gately, p. 21.
14 "cast a spell," Morison, *Christopher Columbus*, p. 12.
15 "as valuable as gold," quoted in Wilford, p. 52.
15 "south with a strong," quoted in Cummins, p. 83.
15 "Halfway between these," quoted in ibid., p. 99.
16 "seem to live," quoted in Morison, *Christopher Columbus*, p. 52.
16 "They invite you," quoted in ibid., p. 51.
16 "The two Christians," quoted in Robicsek, p. 4
17 "No one smoked," Gately, p. 23.
18 "the greatest honor," quoted in Axtell, p. 192.
18 "My two men," quoted in Cummins, p. 115.
20 "as a drug for easing," Klein, Richard, p. 27.
21 "The Floridians," quoted in Morison, *Northern Voyages*, p. 428.
21 "Up comes brave Hawkins," quoted in Bain, p. 22.
22 "His customary cartwheel ruff" and "a dagger," Milton, p. 42.
23 "though ye have had," quoted in Murray, p. 101.
23 "someone who shared," Johnson, *Elizabeth I*, p. 219.
23 "new plush cloak" and "a plashy place," quoted in Lacey, p. 42.
24 "Always keen," Lacey, p. 40.
24 "a metal stopper," Brooks, *Library*, p. 8.
24 "the average gallant," Gately, p. 47.
25 "at bull-baiting," Heimann, p. 40.
26 "There were supposed to be," Brooks, *Library*, p. 6.
26 "This day," quoted in Gately, p. 6.
26 "Children were permitted," Ackroyd, p. 359.
27 "the most sovereign," quoted in Gately, p. 48.
27 "divine," quoted in Fairholt, p. 47.
27 "All they that love," quoted in Gately, p. 48.
27 "But he's a frugal man," quoted in ibid., p. 49.
27 "with the introduction," quoted in Brooks, *Mighty Leaf*, p. 64.
29 "has never, even in the golden age," Gately, p. 40.
29 "They say [tobacco]," quoted in Goodman, p. 45.
30 "determined to put," Gately, p. 49.
30 "I have just got hold of," quoted in Brongers, p. 23.
31 "tobacco and the food he saw," quoted in Darnton, p. 33.
31 "a certain Count Herman," quoted in Fairholt, p. 119.
31 "Certainly the German's love" and "To drink a pint," quoted in ibid., p. 122.
32 "Die geen toeback" and "Whoever does not like," quoted in Brongers, p. 64.
32 "The smell of the Dutch Republic" and "In the middle," Schama, p. 189.
33 "Samurai knights," Gately, p. 57.
33 "When our forces," quoted in ibid., p. 58.
33 "cures troubles" and "has an irritating flavor," quoted in ibid., p. 53.
33 "He had horse manure," Durant and Durant, pp. 531–32.
34 "The Prophet," quoted in Bain, p. 24.

Two: The Enemies of Tobacco

page 38 "See what a wicked," quoted in Robicsek, p. 4.
38 "that they fall down" and "feel a pleasure," quoted in Heimann, p. 16.
39 "sovereign and precious," quoted in Gately, p. 48.
39 "roguish tobacco," quoted in Wagner, p. 11.
39 "It is a shocking thing," quoted in Brooks, *Mighty Leaf*, p. 273.
39 "patrimony of many" and "a fantasticall attracter," quoted in Fairholt, p. 75.
39 "it was decreed," Ackroyd, p. 211.
40 "Under pain," quoted in Robicsek, p. 6.
40 "the use of the herb," quoted in Heimann, p. 38.
40 "take snuff or tobacco," quoted in Fairholt, p. 78n.
40 "the revocation was not," Brooks, *Library*, p. 10.
40 "I say … that the use," quoted in Goodman, p. 78.
41 "an unfortunate Turk," Fairholt, p. 79.
42 "Was ever the destruction," Fairholt, pp. 78–79.
43 "was an aged," quoted in Milton, p. 251.
44 "a man of ponderous erudition" and "lectured Englishmen," Tindall, p. 45.
44 "It was said," Milton, p. 251.
44 "He also spent plenty," Murray, p. 92.
44 "his big head," Bain, p. 15.
44 "In quilted doublet," quoted in Fairholt, p. 80.
45 "What honor," "say without blushing," and "makes a kitchen," James I, n.p.
46 "his members," "a generall sluggishnesse," "O omnipotent power," and "a custome lothsome," ibid., n.p.
47 "immediately intensified," Brooks, *Mighty Leaf*, p. 56.
48 "no heiress," Winton, p. 109.
48 "enjoy sweet embraces," quoted in Lacey, p. 147.
48 "with a fierceness," ibid., p. 148.
49 "sheathed in slabs" and "the gold would reflect," Guadalupi and Shugaar, pp. 25–26.
50 "skulking in an obscure cove," Lacey, p. 220.
50 "in pride exceedeth" and "the greatest Lucifer," quoted in Greenblatt, p. 56.
50 "Let me, therefore," quoted in ibid., pp. 113–14.
51 "It seems a miracle," James I, n.p.
52 "For years," Milton, p. 258.
52 "Never was a man," quoted in ibid., p. 258.
52 "macerating forty roots," Barr, p. 201.
52 "great cordial," quoted in ibid., p. 201.
53 "could be called," Lacey, p. 325.
54 "I shall sorrow," quoted in ibid., p. 359.
54 "It was a humbling," ibid., p. 373.
54 "some female persons" and " 'twas well and properly donne," quoted in Fairholt, p. 52n.
55 "This is a sharp medicine," quoted in Lacey, p. 381.
55 "What dost thou fear?" quoted in Winton, p. 342.
55 "We have not another," quoted in Lacey, p. 382.

Three: The Politics of Tobacco

page 56 "to preach and baptize," quoted in Johnson, *History*, p. 23.
56 "But by Elizabeth I's time," Furnas, p. 49.
56 "oddsticks who had," Fisher, p. 18.

56 "Our land, abounding," quoted in Johnson, *History*, p. 24.
57 "Tradesmen, Serving-men," quoted in Amory, p. 32.
57 "miasmic vapours," quoted in Catton, p. 43.
57 "doggs Catts," quoted in Davidson and Lytle, p. 11.
57 "Every man almost laments," quoted in Gelb, p. 25.
57 "disease, sickness," Hatch, p. 10.
58 "then ripped the childe," quoted in Milton, p. 297.
58 "Now whether she," quoted in Hatch, p. 28.
58 "boots, shoes," quoted in Amory, p. 44.
58 "despair moaned," Miers, p. 28.
59 "silkworm buff," Davidson and Lytle, p. 10.
60 "poor and weake," quoted in Heimann," p. 42.
61 "becomes a source" and "dependent upon," Ehwa, p. 140.
61 "Tobacco [is] verie commodious," quoted in Milton, p. 309.
61 "was by far," quoted in Mossiker, p. 174.
62 "this nation's first business," Heimann, p. 5.
63 "virtually everything else," Hume, p. 323.
63 "since they could not eat money," quoted in Brandon et al., p. 166.
64 "but five or six houses," quoted in Hatch, p. 18.
64 "We have with great joy," quoted in Price, p. 187.
64 "The society that was designed," Davidson and Lytle, p. 10.
65 "Wee found the Colony," quoted in ibid., p. 12.
65 "It was only on virgin land," Boorstin, *Americans*, p. 105.
65 "Extermination of the beaver," Sale, p. 291.
66 "a tomahawk or feathered arrow," Mossiker, p. 301.
66 "to misuse and misemploy," quoted in Fairholt, p. 111.
67 "Whereas it is agreed," quoted in Gately, p. 75.
67 "all Liberties, Franchises," quoted in Gelb, p. 25.
68 "As it is atheism," quoted in Wells, p. 639.
68 "plunge their hands," Liss, p. 12.
68 "in these three years," quoted in Heimann, p. 60.
69 "In the 1660s," Goodman, p. 150.
69 "practically the do-all," Wilstach, p. 61.
69 "soe good a reputation," quoted in Tobacco Institute, *South Carolina and Tobacco*, p. 14.
69 "become renowned," quoted in Tobacco Institute, *Pennsylvania and Tobacco*, p. 10.
69 "for the most part," Bradford, p. 82.
69 "greater attention to hemp," Pomfret and Shumway, p. 230.
70 "these debts had become," quoted in Gately, p. 137.
70 "in four years," Brooks, *Mighty Leaf*, pp. 161–62.
70 "a reasonable fraction," McDonald, p. 70.

Four: The Rise of Tobacco

page 71 "Golden Token," quoted in Wagner, p. 18.
71 "a Dutch man of warre" and "there hath been," quoted in Brooks, *Mighty Leaf*, p. 93.
72 "contrary to Law" and "for his insolence," quoted in Amory, p. 75n.
72 "Every master" and "to preach or teach," quoted in Fairholt, pp. 112–13.
73 "their principal token," Klein, Richard, p. 15.
73 "Superb sermons," Brooks, *Mighty Leaf*, pp. 93–94.

73 "Some parishes," quoted in Boorstin, *Americans*, p. 128.
74 "All in all," Wertenbaker, p. 123.
75 "seemed beguiled by," Langguth, p. 42.
76 "a smoldering eye," Cooke, p. 106.
76 "a dignified gentleman," Robert, *Story*, p. 37.
76 "seems to have begun," Johnson, *History*, p. 110.
77 "one of the few men," Langguth, p. 45.
78 "taken captive," Beeman, p. 19.
78 "rapacious harpies," quoted in Brooks, *Mighty Leaf*, p. 169.
78 "thinks the ready road," Langguth, p. 47.
78 "the father of his people," quoted in Meade, p. 5.
79 "lifted Patrick Henry," Brooks, *Mighty Leaf*, p. 168.
80 "No amount of book-learning," Fisher, pp. 84–85.
80 "Even as planters," Berlin, pp. 15–16.
81 "sound, well conditioned," quoted in Robert, *Tobacco Kingdom*, p. 83.
81 "without benefit," ibid., p. 84.
81 "the counterpart," quoted in Simmons, p. 74.
82 "Dining rooms, parlors," Clinton, p. 18.
82 "carpenters, coopers," Wertenbaker, p. 45.
83 "larger and more successful," Boorstin, *Americans*, p. 108.
83 "Carter and his friends," Burns, James MacGregor, p. 256.
84 "The Merchants of England," quoted in Robert, *Tobacco Kingdom*, p. 22.
84 "There was no one," Bailyn, p. 8.
84 "We shou'd wear it," quoted in Pietrusza, p. 11.
85 "the lightest, mildest," Tobacco Institute, *Virginia and Tobacco*, p. 21.
85 "were to Virginia," Heimann, p. 71.
85 "Trained in the management," ibid., p. 72.
85 "You know," quoted in Ellis, *After the Revolution*, p. 88.
85 "slavery was critical," Ambrose, p. 35.
86 "requires not skilled hands," Wertenbaker, p. 23.
86 "the loss of seedlings," McDougall, p. 159.
86 "it was not uncommon," Ciulla, p. 78.
87 "they import so many Negroes," quoted in Gately, p. 110.
88 "The mortality rate," Chernow, p. 19.
88 "According to the 1810 census," Kurlansky, p. 253.
88 "In New England," Schlesinger, Sr., pp. 63–64.
89 "one man will clean," quoted in Livesay, p. 34.
90 "dangerous and rising tyranny," quoted in Bowen, p. 63.
90 "smoke tobacco till you cannot see," quoted in Ferling, p. 65.
91 "The next day," Langguth, p. 183.
91 "the violent and outrageous proceedings," quoted in Bobrick, p. 91.
91 "devoutly to implore," quoted in ibid., p. 93.
92 "the Old South," Cooke, p. 106.
92 "wholly built upon smoke," quoted in Pomfret and Shumway, p. 35.
93 "despair of any relief," quoted in Maier, p. 4.
93 "basic view rested," Draper, p. 280.
94 "Tobacco functions," Klein, Richard, p. 142.
94 "Although a ball," Heimann, p. 73.
94 "If you can't send money," quoted in Wagner, p. 25.
94 "You stink of brandy," quoted in Tate, p. 67.
95 "useless and barbarous injury," quoted in Heimann, p. 73.

Five: Rush to Judgment

page 97 "the leading figure," Burns, Eric, p. 55.
98 "the poor, upon sailors," quoted in Robert, *Story*, p. 100.
98 "with an effect," quoted in Bobrick, p. 186.
98 "One of the usual effects" and "One of the greatest sots," quoted in Robert, *Story*, p. 106.
99 "black, loathsome discharge," quoted in ibid., p. 108.
99 "good for the fits" and "that it was as well," Eliot, p. 190.
99 "I set down again," Twain, p. 16.
99 "came into the world" and "never to smoke," quoted in Cooper, p. 61.
100 "Rum-drinking," "Tobacco and alcohol were Satan's," and "the great-grandparent vices," quoted in Robert, *Story*, p. 107.
100 "Yes—thou poor degraded creature," quoted in Kobler, pp. 42–43.
101 "a cold abstract theory," quoted in ibid., p. 59.
101 "which is to goe," quoted in Heimann, p. 83.
101 "under the age of 20," quoted in ibid., p. 83.
102 "common idlers," Robicsek, p. 105.
102 "The populace were," quoted in Tobacco Institute, *New York and Tobacco*, pp. 8–9.

Six: Ghost, Body, and Soul

page 103 "Wealthy aristocrats and royalty," Rogozinski, p. 45.
104 "might be an honorable man," Langguth, p. 425.
104 "the disdainful disregard," Bobrick, p. 149.
105 "of the topmost rung," Furnas, p. 223.
106 "without the least attempt," quoted in Tate, p. 210.
106 "a black, twisty," Furnas, p. 223.
107 "quietly smoked his cigar," Turgenev, p. 142.
107 "On a cold morning," quoted in Davidoff, p. 90.
107 "Light me another Cuba," quoted in Bain, p. 36.
107 "conversation was as flat," Flaubert, p. 42.
107 "pouted out his lips," ibid., p. 57.
107 "went swinging off" and "As the horseman," Crane, p. 17.
107 "Had there been half a dozen," Cather, p. 52.
108 "You don't mind my cigar," Thackeray, p. 119.
108 "resented regulations," Wineapple, p. 47.
108 "mingled with camphorated chalk," Bain, p. 72.
108 "a fire at one end," quoted in Wagner, p. 31.
109 "They have a fashion," quoted in Brands, *The Age of Gold*, p. 84.
109 "green herb," Brooks, *Mighty Leaf*, p. 212.
109 "primed for conflict," ibid., p. 212.
110 "chewing herbs," Fairholt, p. 11.
110 "a great Powhatan bowl," quoted in Gately, p. 166.
110 "He had vigorous thoughts," Schlesinger, p. 41.
111 "Brawling in barrooms," Davidson and Lytle, p. 92.
111 "too preposterous," "one of the most," and "And so," Cooke, p. 170.
111 "dysentery and pellagra," Masur, p. 128.
112 "the original populist," Will, p. 16.
112 "one of the most notorious scenes," Burstein, p. 173.
112 "Chief Justice John Marshall," Simmons, p. 98.

113 "If movement," quoted in Seldes, p. 39.
113 "a foppish indulgence," Carter, p. 26.
114 "It may be," Brooks, *Mighty Leaf*, p. 214.
114 "Europe was everything," Gabler, p. 26.
115 "he had long been optimistic," Simmons, p. 95.
116 "As the *Britannia*," Kaplan, p. 126.
116 "I can give you no conception," quoted in Simmons, p. 95.
116 "Not even the Beatles's reception," ibid., p. 95.
116 "a select party," Furnas, p. 456.
116 "yellow streams," quoted in Robert, *Story*, p. 103.
117 "You never can conceive," quoted in ibid., p. 103.
117 "expectorate in dreams," quoted in Tate, p. 17.
117 "Set right there," quoted in Wagner, p. 232.
117 "even steady old chewers," quoted in ibid., p. 232.
117 "Washington may be called," quoted in Brooks, *Mighty Leaf*, pp. 215–16.
118 "the most sickening," quoted in Simmons, p. 117.
118 "Desolation was the only feeling," quoted in ibid., pp. 25–26.
118 "the gentlemen spit," quoted in Simmons, p. 33.
118 "I may not describe," Trollope, pp. 15–16.
119 "This has been called," ibid., pp. 58–59n.
119 "smoke, chew, spit," quoted in Pierson, p. 478.
119 "No single other thing," Furnas, p. 243.
119 "Snuffing is through," quoted in Curtis, p. 22.
120 "heals colds," quoted in Goodman, p. 79.
121 "found to occasion," Fairholt, p. 242.
121 "The mineral, vegetable," Curtis, pp. 84–85.
122 "One has boxes," quoted in Goodman, p. 74.
123 "If we supposed," quoted in Fairholt, p. 285.
123 "that their sense," Handlin, p. 102.
123 "superfluous humours," quoted in Brooks, *Mighty Leaf*, p. 123.
124 "the subject of more," ibid., p. 122.
124 "a cure for the Hickups," quoted in Boorstin, *Seekers*, p. 44.
124 "The female snuff-dipper," quoted in Rogozinski, p. 43.
125 "(1) absence of family restraints" and "did a large part," Robert, *Story*, p. 119.
125 "Doctor, can you tell me," quoted in Kobler, p. 65.
125 "No caricaturist," Bain, p. 71.

Seven: The Cigarette

page 129 "a poor man's by-product," Heimann, p. 203.
130 "ascending the social ladder," Klein, Richard, p. 204.
130 "The general professional incompetence," Reader, p. 40.
130 "cylinders of straw-coloured paper," Wilson, p. 197.
130 "A paper cigarette," quoted in Sobel, p. 9.
131 "war and advertising," Sobel, p. 14.
131 "When one smokes," Klein, Richard, pp. 139–41.
132 "The decadence of Spain," quoted in Wagner, p. 41.
132 "In 1854," Tennant, p. 15.
132 "produced a sense of guilt," Sobel, p. 13.
132 "A woman smoking," Klein, Richard, p. 130.
134 "poured a flow," Hirschfelder, p. 12.

134 "Bonsack's distinctive contributions," Tennant, p. 18.
134 "On April 30, 1884," Hirschfelder, p. 13.
135 "These cigarettes," quoted in ibid., p. 13.
136 "scores of railroads," Cooper, pp. 24–25.
137 "In fact, the ingestion," Sobel, p. 69.
137 "Phosphorus fumes," Wilson, p. 511.
137 "Farmers who left the countryside," Sobel, p. 42.
138 "was drugged with opium," Brooks, *Mighty Leaf*, pp. 253–54.
139 "was a function of space," Heimann, p. 142.
139 "It has become impossible," quoted in ibid., p. 237.

Eight: The Carry Nation of Tobacco

page 140 "tall, ungainly, and rather bony," Sobel, p. 53.
140 "little white slavers," quoted in Heimann, p. 215.
141 "as liquor impaired," Kobler, pp. 142–43.
141 "leads to the devil" and "no man would ever be seen," quoted in Tate, p. 28.
141 "temperance work was already," Warfield, p. 244.
142 "Oh, the deadly cigarette," Nation, p. 350.
142 "had not his blood," quoted in Holbrook, p. 104.
142 "With phenomenal restraint," Warfield, p. 245.
142 "Yours for the extermination," quoted in Tate, p. 62.
142 "one of my prize-fighting friends," quoted in Robert, *Story*, p. 172n.
143 "The W.C.T.U.," Dalton, p. 251.
143 "The White House guard," Holbrook, p. 104.
144 "destroyed red corpuscles," Warfield, p. 244.
144 "very sick" and "to be hanged," quoted in Sobel, p. 54.
144 "jumped from a three-story window," quoted in Warfield, p. 244.
144 "cigarette face," quoted in Sobel, p. 53.
144 "took to drink," ibid., p. 53.
144 "Thousands of clear-eyed," quoted in Tate, p. 40.
145 "which advised America's youth," Gately, p. 26.
145 "A gaunt, middle-aged woman," Warfield, p. 246.
146 "'Horrible, Horrible," quoted in Dedmon, p. 308.
146 "cigarette smokers are men," quoted in Sobel, p. 61.
146 "Boys who smoke cigarettes," quoted in ibid., p. 61.
146 "tends to insanity," "is powerful," and "arrests the growth," quoted in Tate, p. 52.
147 "was convinced," quoted in Sobel, p. 87.
147 "If all boys," quoted in Gately, p. 231.
147 "The injurious agent," quoted in Brinkley, p. 165.
147 "Smoke cigarettes?" quoted in Brooks, *Mighty Leaf*, p. 259.
148 "tramps and ragpickers," ibid., p. 253.
148 "It's the Dutchmen," quoted in ibid., p. 259.
149 "illegal for any person" and "the manufacture, sale," Wagner, p. 55.
149 "flagrantly and openly violated," Tate, p. 61.
149 "The smoking of cigarettes," quoted in Tennant, pp. 133–34.
149 "inferior breeds of people," Tate, p. 18.
150 "evil of great magnitude" and "to labor in a spirit," quoted in Robert, *Story*, p. 170.
150 "Will you wholly abstain," quoted in ibid., p. 171.
150 "after the present session," quoted in Tennant, p. 132.
151 "patented a mouthwash," Sobel, p. 55.

151 "Messenger boys," Tate, p. 58.
151 "produce extreme nausea," ibid., p. 58.
152 "the only guaranteed," Shenkman and Reiger, p. 165.
152 "Remember gentian root," quoted in Warfield, p. 247.
153 "ladies and gentlemen," Furnas, p. 140.
153 "It is difficult," ibid., p. 457.
154 "You would never know," quoted in Neville-Sington, p. 133.
154 "The city stepped up," Larson, p. 212.
155 "The Sunday papers," Furnas, p. 908.
157 "You ask me what," quoted in Sobel, p. 84.
157 "A cigarette may make," quoted in Tate, p. 71.
157 "a means of diversion," quoted in ibid., p. 70.
158 "Coffin nails," quoted in ibid., p. 84.
158 "Sophie Tucker," ibid., p. 83.
158 "There are hundreds of thousands," quoted in ibid., p. 77.
158 "actually encouraged soldiers," ibid., p. 77.
159 "one of the greatest blessings," quoted in ibid., p. 81.
159 "Any man in uniform," Sobel, p. 86.
159 "the British infantryman's tobacco ration," Gately, pp. 231–32.
160 "something almost inspiring," quoted in Burnham, p. 83.
160 "But the helmet," Sobel, p. 83.
160 "Conversely, the worst moments," Klein, Richard, p. 148.
161 "My chest was splashed," quoted in Gately, pp. 235–36.
161 "Americans who followed," Sobel, p. 85.
162 "People seem to be," quoted in Tate, p. 62.
162 "ten cigars, twenty cigarettes" and "decent," Remarque, p. 4.
162 "Cigarettes and pipes out," ibid., p. 40.
164 "Prohibition is won," quoted in Heimann, p. 250.
164 "In one tenement," Burns and Dunn, p. 31.
166 "We must not let," Lord, p. 5.
166 "I think it is fine," quoted in Sobel, p. 89.
166 "With no regular salary," Tate, p. 131.
166 "We are out to put," quoted in ibid., p. 131.
166 "harangued women smokers," ibid., p. 131.
167 "fine character" and "Haven't you a little admiration," quoted in ibid., p. 132.

Nine: The Last Good Time

page 169 "the seven fat years," ibid., p. 28.
169 "into regions once considered," ibid., p. 3
170 "to piece out pillows," Wagner, p. 55.
170 "Ask Dad, He Knows," quoted in Brooks, *Mighty Leaf*, p. 261.
170 "Modern business," quoted in Klein, Maury, p. 109.
170 "The product itself," quoted in Shorris, p. 51.
172 "I won't be played," quoted in Gately, p. 98.
173 "The cigarette smoking woman," Furnas, pp. 894–95.
173 "Ladies may," quoted in Robert, *Story*, p. 252.
174 "for impairing the morals," quoted in Tate, p. 112.
174 "cigarete [*sic*] fiend," quoted in ibid., p. 112.
174 "a lady cowering," Lord, p. 274.
174 "You can't do that," quoted in ibid., p. 274.

174 "The women can not save," quoted in Tate, p. 112.
174 "muscling in on," O'Hara, p. 8.
174 "now strewed the dinner table," Allen, p. 110.
175 "More women now" and "Smoking is a sublimation," quoted in Sobel, p. 95.
175 "Particularly when smoked," Tate, p. 24.
175 "And tell me, Niel," Cather, pp. 111–12.
176 "is a dirty, expensive" and "To which I am devoted," quoted in Robert, *Story*, p. 254.
176 "Women—when they smoke at all" and "a soothing blend," quoted in Gately, p. 244.
177 "rawboned, diminutive figure," Tye, p. 27.
177 "behind a desk," ibid., pp. 27–28.
177 "radios in every room," Wagner, p. 71.
178 "dominion and control," quoted in Gately, pp. 222–23.
178 "looked at the corner," quoted in Wagner, p. 57.
179 "to ask other photographers," Tye, p. 24.
180 "directly to change," ibid., p. 25.
181 "Do not let anyone," quoted in Wagner, p. 58.
181 "Lucky is a marvelous pal," quoted in ibid., p. 58.
181 "representing some two dozen," ibid., p. 59.
181 "I Light a Lucky," quoted in Sobel, p. 100.
181 "Sweets 'fixed' saliva," ibid., p. 100.
181 "A reasonable proportion," quoted in Wagner, pp. 58–59.
181 "Don't neglect your candy ration!" quoted in ibid., p. 59.
182 "Eat a Chocolate," quoted in ibid., p. 60.
182 "I've spent millions," quoted in Tye, p. 38.
183 "green gloves and green shoes," ibid., p. 39.
183 "invited fashion editors," ibid., p. 39.
183 "from the first strawberry flush," ibid., p. 40.
184 "Bernays emphasized," quoted in ibid., p. 98.
184 "a little distress selling," quoted in Gordon and Gordon, p. 88.
185 "Emaciated children," quoted in Kennedy, p. xiv.
185 "butt pickers," Okrent, p. 405.
186 "sharecropper with a wife," Robert, *Story*, p. 207.
187 "could not afford," Gately, p. 247.
187 "Early in the depression," Sobel, p. 112.
188 "The way a man," Gately, p. 267.

Ten: The Case Against Tobacco

page 189 "makes a kitchen," James I, n.p.
189 "The scientific study," Wagner, p. 64.
190 "an age when," Neville-Sington, p. 351.
191 "the disastrous effects," quoted in Tate, p. 19.
191 "thoroughly ill," James, p. 334.
191 "the most common cause," Wagner, p. 42.
192 "able to produce cancerous tumors," ibid., p. 68.
193 "Smoking is associated," quoted in ibid., p. 69.
193 "two-thirds of the non-smokers," ibid., p. 69.
193 "presents the world's first" and "the usual list," Proctor, p. 194.
193 "Was the deceased," quoted in ibid., p. 195.
194 "arguably the most comprehensive," ibid., p. 184.
194 "smoke alley," quoted in ibid., p. 184.

194 "lips, tongue," ibid., pp. 184–85.
197 "But he had another job," Cramer, p. 291.
198 "The entire amount," quoted in Heimann, p. 242.
199 "lacked the 'cleanliness,'" Sobel, p. 132.
200 "The implication was," ibid., p. 132.
200 "The soldiers in worrying," quoted in Robert, *Story*, p. 270.
200 "The boys in the fox-holes," quoted in ibid., pp. 269–70.
201 "From the soldiers' point of view," Gately, p. 260.
202 "how his hands shook," ibid., p. 266.
203 "accounted for 1.4%," Tennant, p. 3.
203 "World War II had ended," Patterson, p. 1996.
205 "While Secretary of State," Sobel, p. 145.
205 "possessed all the basic qualities," ibid., p. 141.
206 "for two years after V-E Day," Heimann, p. 243.
206 "cautious hedonism," Sobel, p. 148.
207 "a death watch" and "the tobacco industry," quoted in ibid., p. 167.
207 "the tremendous and unprecedented," quoted in Wagner, p. 76.
207 "The death rate," ibid., p. 77.
208 "The tobacco industry," ibid., p. 78.
208 "Before You Smoke," quoted in ibid., p. 124.
209 "reach[ed] a circulation" and "RECENT REPORTS," quoted in Gately, p. 287.
210 "a scientist," quoted in Gately, p. 288.
211 "no one was to be given," Oakley, p. 105.
211 "George Washington Hill," quoted in Sobel, p. 182.
212 "NO MATCHES REQUIRED" and "No Nicotine," quoted in Tate, p. 19.
212 "30,000 filaments," "40,000 filter traps," and "Miracle Tip," quoted in Wagner, p. 90.
214 "Truth is you cain't buy," quoted in Heat Moon, p. 54.

Eleven: The Turning Point

page 218 "few Americans knew," Margolis, p. 96.
218 "That matter is sensitive enough," quoted in Hunter, p. 65.
219 "We were scared," quoted in ibid., p. 65.
220 "Cigarette smoking" and following quotes are taken from Surgeon General's Advisory Committee on Smoking and Health.
221 "Tobacco amblyopia" and following quotes, ibid.
222 "Some puffed determinedly," Hunter, p. 65.
222 "Speaking as a doctor," quoted in "Cigarette Smoking Is a Health Hazard," p. 48.
223 "I endorse wholeheartedly," quoted in Allan, p. 65.
224 "simply evaluated and re-processed" and "not really the last word," quoted in ibid., p. 65.
224 "Staff members at these hospitals," Wagner, p. 132.
225 "On top of Old Smoking," quoted in ibid., p. 148.
225 "about three packs a day," Caro, p. 494.
225 "fingers were stained," ibid., p. 617.
225 "I'd rather have," quoted in ibid., p. 624.
225 "found the best filter yet," quoted in Gately, pp. 296–97.
229 "For instance," quoted in Wagner, p. 221.
230 "I guarantee," quoted in Gately, p. 336.
230 "At least one company," Lyman, p. E1.
230 "Dozens of theaters," Tate, p. 104.

231 "the public health," quoted in Gately, p. 310.
232 "The Justice Department," Lichtblau, p. A1.
232 "During a period," Weinstein.
233 "pathetic," "freedom of commercial expression," and "gives the press," quoted in "Cigarette Ad Ban."
234 "The modern cigarette," quoted in Goodman, p. 6.
234 "Even if there is," Shaw, p. 131.
237 "the companies have deceived," Janofsky.
237 "The core theme," quoted in ibid.
238 "represents escape," Goodman, p. 113.
238 "I've spent the last month," transcribed by the author from the BBC documentary *The Tobacco Wars.*
239 "an anti-social action," quoted in Tye, p. 49.
239 "enlisting moviemakers," quoted in ibid., p. 49.
239 "got rid of a sense of guilt," quoted in ibid., p. 49.

Epilogue: "The Ten O'Clock People"

All quotes in this section are from King, pp. 501–58.

SELECT BIBLIOGRAPHY

Ackroyd, Peter. *London: The Biography*. New York: Doubleday, 2000.
Allan, John H. "Tobacco Institute Says Report 'Is Not Final Chapter' in Debate over Health." *New York Times*, January 12, 1964.
Allen, Frederick Lewis. *Only Yesterday: An Informal History of the Nineteen-Twenties*. New York: Harper and Brothers, 1957.
Ambrose, Stephen. *Undaunted Courage: Meriwether Lewis, Thomas Jefferson, and the Opening of the American West*. New York: Simon & Schuster, 1996.
Amory, Cleveland. *Who Killed Society?* New York: Harper and Brothers, 1960.
Anthony, Katharine. *Queen Elizabeth*. New York: Literary Guild, 1929.
Axtell, James. *Encounters in Colonial North America*. New York: Oxford University Press, 1992.
Bailyn, Bernard. *To Begin the World Anew: The Genius and Ambiguities of the American Founders*. New York: Knopf, 2003.
Bain, John, Jr. *Tobacco in Song and Story*. New York: Caldwell, 1896.
Barr, Andrew. *Drink: A Social History of America*. New York: Carroll and Graf, 1999.
Barth, Ilene. *The Smoking Life*. Columbus, Miss.: Genesis, 1997.
Barzun, Jacques. *From Dawn to Decadence: 1500 to the Present, 500 Years of Western Cultural Life*. New York: HarperCollins, 2000.
Beaumier, John Paul, and Lewis Camp. *The Pipe Smoker*. New York: Harper and Row, 1980.
Beeman, Richard R. *Patrick Henry: A Biography*. New York: McGraw-Hill, 1974.
Beran, Michael Knox. *Jefferson's Demons: Portrait of a Restless Mind*. New York: Free Press, 2003.
Berlin, Ira. *Many Thousands Gone: The First Two Centuries of Slavery in North America*. Cambridge, Mass.: Harvard University Press, 1998.
Bobrick, Benson. *Angel in the Whirlwind: The Triumph of the American Revolution*. New York: Simon and Schuster, 1997.
Boorstin, Daniel J. *The Americans: The Colonial Experience*. New York: Random House, 1958.
———. *The Discoverers*. New York: Random House, 1983.
———. *The Seekers*. New York: Random House, 1983.
Bowen, Catherine Drinker. *Yankee from Olympus: Justice Holmes and His Family*. Norwalk, Conn.: Easton Press, 1990
Bradford, William. *Of Plymouth Plantation: 1620–1647*. New York: Alfred A. Knopf, 1970.
Bramhall, William. *The Great American Misfit*. New York: Clarkson N. Potter, 1982.
Brandon, William, et al. *The American Heritage Book of Indians*. New York: Simon and Schuster, 1961.

Brands, H. W. *The Age of Gold: The California Gold Rush and the New American Dream.* New York: Doubleday, 2002.

———. *The First American: The Life and Times of Benjamin Franklin.* New York: Doubleday, 2000.

Braudel, Fernand. *The Structures of Everyday Life: Civilization and Capitalism, 15th–18th Century*, vol. 1. New York: Harper and Row, 1981.

Bridenbaugh, Carl. *Jamestown: 1544–1699.* New York: Oxford University Press, 1980.

Brinkley, Douglas. *Wheels for the World: Henry Ford, His Company, and a Century of Progress.* New York: Viking, 2003.

Brongers, Georg. *Nicotiana Tabacum.* Groningen: Theodorus Niemeyer, 1964.

Brookhiser, Richard. *America's First Dynasty: The Adamses, 1735–1918.* New York: Free Press, 2002.

Brooks, Jerome. *The Library Relating to Tobacco, Collected by George Arents.* New York: New York Public Library, 1944.

———. *The Mighty Leaf: Tobacco through the Centuries.* Boston: Little, Brown, 1952.

Burnham, John C. *Bad Habits: Drinking, Smoking, Taking Drugs, Gambling, Sexual Misbehavior, and Swearing in American History.* New York: New York University Press, 1993.

Burns, Eric. *The Spirits of America: A Social History of Alcohol.* Philadelphia: Temple University Press, 2003.

Burns, James MacGregor. *The Vineyard of Liberty.* New York: Alfred A. Knopf, 1982.

Burns, James MacGregor, and Susan Dunn. *The Three Roosevelts: Patrician Leaders Who Transformed America.* New York: Atlantic Monthly Press, 2001.

Burstein, Andrew. *The Passions of Andrew Jackson.* New York: Alfred A. Knopf, 2003.

Caro, Robert. *The Years of Lyndon Johnson: The Path to Power.* New York: Alfred A. Knopf, 1982.

Carter, Stephen L. *Civility: Manners, Morals and the Etiquette of Democracy.* New York: Basic Books, 1998.

Cather, Willa. *A Lost Lady.* New York: Alfred A. Knopf, 1978.

Catton, Bruce. *Reflections on the Civil War.* New York: Doubleday, 1981.

Chernow, Ron. *Alexander Hamilton.* New York: Penguin Press, 2004.

"Cigarette Ad Ban," Associated Press, April 28, 1999.

"Cigarette Smoking Is a Health Hazard." *Newsweek*, January 20, 1964.

Ciulla, Joanne B. *The Working Life: The Promise and Betrayal of Modern Work.* New York: Times Books, 2000.

Clinton, Catherine. *The Plantation Mistress: Woman's World in the Old South.* New York: Pantheon, 1982.

Cooke, Alistair. *America.* New York: Alfred A. Knopf, 1973.

Cooper, Robert. *Around the World with Mark Twain.* New York: Arcade, 2000.

Cramer, Richard Ben. *Joe DiMaggio: The Hero's Life.* New York: Simon and Schuster, 2000.

Crane, Stephen. *The Red Badge of Courage.* New York: Heritage Press, 1944.

Cummins, John, trans. *The Voyage of Christopher Columbus: Columbus's Own Journal of Discovery.* New York: St. Martin's Press, 1992.

Curtis, Mattoon M. *The Story of Snuff and Snuff Boxes.* New York: Liveright, 1935.

Dalton, Kathleen. *Theodore Roosevelt: A Strenuous Life.* New York: Alfred A. Knopf, 2002.

Darnton, Robert. *The Great Cat Massacre and Other Episodes in French Cultural History.* New York: Basic Books, 1984.

Davidoff, Zino. *The Connoisseur's Book of the Cigar.* New York: McGraw-Hill, 1967.

Davidson, James West, and Mark Hamilton Lytle. *After the Fact: The Art of Historical Detection.* New York: Alfred A. Knopf, 1982.

Dedmon, Emmett. *Fabulous Chicago.* New York: Atheneum, 1981.

Diaz de Castillo, Bernal. *The Conquest of New Spain.* Middlesex: Penguin, 1963.

Draper, Theodore. *A Struggle for Power: The American Revolution.* New York: Times Books, 1996.

Durant, Will, and Ariel Durant. *The Story of Civilization, Volume 7: The Age of Reason Begins.* New York: Simon and Schuster, 1961.
Ehwa, Carl, Jr. *The Book of Pipes and Tobacco.* New York: Random House, 1973.
Eliot, George. *Silas Marner.* Norwalk, Conn.: Easton Press, n.d.
Ellis, Joseph J. *After the Revolution: Profiles of Early American Culture.* New York: W. W. Norton, 2002.
———. *American Sphinx: The Character of Thomas Jefferson.* New York: Alfred A. Knopf, 1997.
Erickson, Arvel B., and Martin J. Havran. *England: Prehistory to Present.* New York: Praeger, 1968.
Fairholt, F. W. *Tobacco: Its History and Associations.* Detroit: Singing Tree, 1968.
Ferling, John. *A Leap in the Dark: The Struggle to Create the American Republic.* Oxford: Oxford University Press, 2003.
Fisher, Sydney George. *Men, Women and Manners in Colonial Times*, vol. 2. Philadelphia: J. B . Lippincott, 1897.
Flaubert, Gustave. *Madame Bovary.* Norwalk, Conn.: Easton Press, 1978.
Forbes, Esther. *Paul Revere and the World He Lived In.* New York: Houghton Mifflin, 1972.
Fox, Stephen. *The Mirror Makers: A History of American Advertising and Its Creators.* New York: William Morrow, 1984.
Frazer, Sir James George. *The Golden Bough: A Study in Magic and Religion.* New York: Macmillan, 1979.
Frum, David. *How We Got Here: The 70s, the Decade That Brought You Modern Life (for Better or Worse).* New York: Basic Books, 2000.
Furnas, J. C. *The Americans: A Social History of the United States.* New York: G. P. Putnam's Sons, 1969.
Gabler, Neal. *Life: The Movie.* New York: Alfred A. Knopf, 1999.
Gately, Iain. *Tobacco: A Cultural History of How an Exotic Plant Seduced Civilization.* New York: Grove Press, 2001.
Gelb, Norman. *Less than Glory: A Revisionist's View of the American Revolution.* New York: G. P. Putnam's Sons, 1984.
Goodman, Jordan. *Tobacco in History: The Cultures of Dependence.* London: Routledge, 1994.
Gordon, Lois, and Alan Gordon. *American Chronicle: Six Decades in America Life, 1920–1980.* New York: Atheneum, 1987.
Grant, Jane. *Ross,* The New Yorker *and Me.* New York: Reynal, 1968.
Greenblatt, Stephen J. *Sir Walter Ralegh: The Renaissance Man and His Roles.* New Haven, Conn.: Yale University Press, 1973.
Guadalupi, Gianni, and Antony Shugaar. *Latitude Zero: Tales of the Equator.* New York: Carroll and Graf, 2001.
Halliday, E. M. *Understanding Thomas Jefferson.* New York: HarperCollins, 2001.
Hallahan, William H. *The Day the American Revolution Began: 19 April 1775.* New York: William Morrow, 2000.
Handlin, Oscar. *A Restless People: Americans in Rebellion, 1770–1787.* Garden City, N.Y.: Anchor Press/Doubleday, 1982.
Hatch, Charles E., Jr. *The First Seventeen Years: Virginia, 1607–1624.* Charlottesville: University of Virginia Press, 1957.
Heat Moon, William Least. *Blue Highways: A Journey into America.* Boston: Atlantic-Little-Brown, 1982.
Heimann, Robert K. *Tobacco and Americans.* New York: McGraw-Hill, 1960.
Herment, Georges. *The Pipe.* New York: Simon and Schuster: 1955.
Hiebert, Ray Eldon. *Courtier to the Crowd: The Story of Ivy Lee and the Development of Public Relations.* Ames: Iowa State University Press, 1966.
Hirschfelder, Arlene. *Kick Butts: A Kid's Action Guide to a Tobacco-Free America.* Parsippany, N.J.: Julian Messenger, 1998.

Holbrook, Stewart H. *Dreamers of the American Dream.* New York: Doubleday, 1981.
Hume, Ivor Noel. *The Virginia Adventure.* New York: Alfred A. Knopf, 1994.
Hunter, Marjorie. "Smoking Banned at News Parley." *New York Times,* January 12, 1964.
Irwin, Margaret. *That Great Lucifer.* New York: Harcourt, Brace, 1960.
James, Henry. *The Portrait of a Lady.* Norwalk, Conn.: Easton Press, 1978.
James I. *A Counterblaste to Tobacco.* Amsterdam: Da Capo Press, 1969.
Janofsky, Michael. "Lawyers Start Their Defense of Tobacco." *New York Times,* September 23, 2004.
Johnson, Paul. *Elizabeth I.* New York: Holt, Rinehart and Winston, 1974.
———. *A History of the American People.* New York: HarperCollins, 1997.
Jones, Landon Y. *Great Expectations: America and the Baby Boom Generation.* New York: Coward, McCann and Geoghegan, 1980.
Kaplan, Fred. *Dickens: A Biography.* New York: William Morrow, 1988.
Keesecker, William F., ed. *A Calvin Treasury.* New York: Harper Brothers, 1961.
Kennedy, David M. *Freedom from Fear: The American People in Depression and War, 1929–1945.* New York: Oxford University Press, 1999.
Kerr, Daisy. *Keeping Clean: A Very Peculiar History.* New York: Franklin Watts, 1995.
King, Stephen. *Nightmares and Dreamscapes.* New York: Viking, 1993.
Klein, Maury. *Rainbow's End: The Crash of 1929.* New York: Oxford University Press, 2001.
Klein, Richard. *Cigarettes Are Sublime.* Durham, N.C.: Duke University Press, 1993.
Kobler, John. *Ardent Spirits: The Rise and Fall of Prohibition.* New York: G. P. Putnam's Sons, 1973.
Kurlansky, Mark. *Salt: A World History.* New York: Walker and Company, 2002.
Lacey, Robert. *Sir Walter Ralegh.* New York: Atheneum, 1974.
Langguth, A. J. *Patriots: The Men Who Started the American Revolution.* New York: Simon and Schuster, 1988.
Larson, Erik. *The Devil in the White City: Murder, Magic, and Madness at the Fair That Changed America.* New York: Crown, 2003.
Lichtblau, Eric. "U.S. Lawsuit Seeks Tobacco Profits." *New York Times,* March 18, 2003.
Liss, David. *A Conspiracy of Paper.* New York: Random House, 2000.
Livesay, Harold C. *American Made: Men Who Shaped the American Economy.* Boston: Little, Brown, 1979.
Longfellow, Henry Wadsworth. "The Song of Hiawatha." In *Longfellow: Poems and Other Writings.* New York: Library of America, 2000.
Lord, Walter. *The Good Years: From 1900 to the First World War.* New York: Harper and Brothers, 1960.
Lyman, Rick. "In the 80s: Lights ! Camera! Cigarettes!" *New York Times,* March 12, 2002.
Maier, Pauline. *From Resistance to Revolution: Colonial Radicals and the Development of American Opposition to Britain, 1765–1776.* New York: Alfred A. Knopf, 1972.
Manchester, William. *The Last Lion, Winston Spencer Churchill: Visions of Glory, 1874–1932.* Boston: Little, Brown, 1983.
Margolis, Jon. *The Last Innocent Year: America in 1964: The Beginning of the "Sixties."* New York: Morrow, 1999.
Masur, Louis P. *1831: Year of Eclipse.* New York: Hill and Wang, 2001.
McDonald, Forest. *E Pluribus Unum: The Formation of the American Republic, 1776–1790.* Boston: Houghton Mifflin, 1965.
McDougall, Walter A. *Freedom Just around the Corner: A New American History, 1585–1828.* New York: HarperCollins, 2004.
Meade, Robert Douthat. *Patrick Henry: Patriot in the Making.* Philadelphia: J. B. Lippincott, 1957.
Miers, Earl Schenck. *Blood of Freedom: The Story of Jamestown, Williamsburg, and Yorktown.* Williamsburg, Va.: Colonial Williamsburg, 1958.
Miller, Douglas T., and Marion Nowak. *The Fifties: The Way We Really Were.* Garden City, N.Y.: Doubleday, 1977.

Mills, Lewis Sprague. *The Story of Connecticut.* New York: Charles Scribner's Sons, 1932.
Milton, Giles. *Big Chief Elizabeth: The Adventures and Fate of the First English Colonists in America.* New York: Farrar, Straus and Giroux, 2000.
Morison, Samuel Eliot. *Christopher Columbus, Mariner.* Boston: Little, Brown, 1935.
———. *The European Discovery of America: The Northern Voyages, A.D. 500–1600.* New York: Oxford University Press, 1971.
———. *The European Discovery of America: The Southern Voyages, A.D. 1492–1616.* New York: Oxford University Press, 1974.
Mossiker, Frances. *Pocahontas: The Life and the Legend.* New York: Alfred A. Knopf, 1976.
Murray, Jane. *The Kings and Queens of England: A Tourist Guide.* New York: Scribner's, 1974.
Nation, Carry A. *The Use and Need of the Life of Carry A. Nation.* Topeka, Kan.: F. M. Steves and Sons, 1905.
Neville-Sington, Pamela. *Fanny Trollope: The Life and Adventures of a Clever Woman.* New York: Viking, 1997.
Oakley, J. Ronald. *God's Country: America in the Fifties.* New York: Dembner Books, 1986.
Okrent, Daniel. *Great Fortune: The Epic of Rockefeller Center.* New York: Viking, 2003.
O'Hara, John. *Appointment in Samarra.* New York: Vintage Books, 1991.
Olshavsky, Richard W. *No More Butts: A Psychologist's Approach to Quitting Cigarettes.* Bloomington: Indiana University Press, 1977.
Osgood, Herbert L. *The American Colonies in the Seventeenth Century,* vol. 3. New York: Macmillan, 1907.
Patterson, James T. *Grand Expectations: The United States, 1945–1971.* New York: Oxford University Press, 1996.
Pierson, George Wilson. *Tocqueville in America.* Baltimore: Johns Hopkins University Press, 1996.
Pietrusza, David. *Smoking.* San Diego, Calif.: Lucent Press, 1973.
Pomfret, John E., and Floyd M. Shumway. *Founding the American Colonies: 1583–1660.* New York: Harper and Row, 1970.
Porter, Roy. *The Greatest Benefit to Mankind: A Medical History of Humanity.* New York: W.W. Norton, 1998.
Price, David A. *Love and Hate in Jamestown: John Smith, Pocahontas, and the Heart of a New Nation.* New York: Alfred A. Knopf, 2003.
Priestley, Herbert Ingram. *The Coming of the White Man.* New York: Macmillan, 1929.
Pringle, Laurence. *Smoking: A Risky Business.* New York: William Morrow, 1996.
Proctor, Robert N. *The Nazi War on Cancer.* Princeton, N.J.: Princeton University Press, 1999.
Randall, Willard Sterne. *Thomas Jefferson: A Life.* New York: Henry Holt, 1993.
Reader, W. J. *Victorian England.* New York: G. P. Putnam's Sons, 1973.
Reader's Digest Association. *America's Fascinating Indian Heritage.* Pleasantville, N.Y.: Reader's Digest Association, 1978.
Remarque, Erich Maria. *All Quiet on the Western Front.* Norwalk, Conn.: Easton Press, 1969
Rice, C. Duncan. *The Rise and Fall of Black Slavery.* New York: Harper and Row, 1975.
Robert, Joseph C. *The Story of Tobacco in America.* New York: Alfred A. Knopf, 1949.
———. *Tobacco Kingdom.* Gloucester, Mass.: Peter Smith, 1965.
Robicsek, Francis. *The Smoking Gods: Tobacco in Maya Art, History and Religion.* Norman: University of Oklahoma Press, 1978.
Rogozinski, Jan. *Smokeless Tobacco in the Western World, 1550–1950.* New York: Praeger, 1990.
Rorabaugh, W. J. *The Alcoholic Republic: An American Tradition.* New York: Oxford University Press, 1979.
Royster, Charles. *The Fabulous History of the Dismal Swamp Company.* New York: Alfred A. Knopf, 1999.
Sale, Kirkpatrick. *The Conquest of Paradise: Christopher Columbus and the Columbian Legacy.* New York: Alfred A. Knopf, 1990.

Sawyer, Joseph Dillaway. *History of the Pilgrims and Puritans*, vol. 2. New York: Century History, 1922.
Sauer, Carl O. *Seventeenth Century North America*. Berkeley, Calif.: Turtle Island Foundation, 1980.
Schama, Simon. *The Embarrassment of Riches: An Interpretation of Dutch Culture in the Golden Age*. New York: Alfred A. Knopf, 1987.
Schlesinger, Arthur M., Jr. *The Age of Jackson*. Boston: Little, Brown, 1945.
Schlesinger, Arthur M., Sr. *The Birth of the Nation: A Portrait of the American People on the Eve of Independence*. New York: Alfred A. Knopf, 1969.
Seldes, Gilbert. *The Stammering Century*. Gloucester, Mass.: Peter Smith, 1972.
Shaw, David. *The Pleasure Police: How Bluenose Busybodies and Lily-Livered Alarmists Are Taking All the Fun Out of Life*. New York: Doubleday, 1996.
Shenkman, Richard, and Kurt Reiger. *One-Night Stands with American History*. New York: William Morrow, 1980.
Shorris, Earl. *A Nation of Salesmen*. New York: W. W. Norton, 1994.
Simmons, James C. *Star-Spangled Eden: Nineteenth-Century America through the Eyes of Dickens, Wilde, Frances Trollope, Frank Harris, and Other British Travelers*. New York: Carroll and Graf, 2000.
Sobel, Robert. *They Satisfy: The Cigarette in American Life*. New York: Doubleday, 1978.
Surgeon General's Advisory Committee on Smoking and Health. *Smoking and Health: Report of the Advisory Committee to the Surgeon General of the Public Health Service*. Public Health Service Publication 1103. Washington, D.C.: U.S. Government Printing Office, 1964.
Tate, Cassandra. *Cigarette Wars*. New York: Oxford University Press, 1998.
Taylor, Robert Lewis. *Vessel of Wrath: The Life and Times of Carry Nation*. New York: New American Library, 1966.
Tennant, Richard B. *The American Cigarette Industry: A Study in Economic Analysis and Public Policy*. New Haven, Conn.: Yale University Press, 1950.
Thackeray, William Makepeace. *Vanity Fair*. Norwalk, Conn.: Easton Press, 1979.
Tindall, George Brown. *America: A Narrative History*. New York: W. W. Norton, 1984.
Tobacco Institute. *Connecticut and Tobacco* [pamphlet]. Washington, D.C., 1972.
———. *New York and Tobacco* [pamphlet]. Washington, D.C., 1971.
———. *Pennsylvania and Tobacco* [pamphlet]. Washington, D.C., 1972.
———. *South Carolina and Tobacco* [pamphlet]. Washington, D.C., 1972.
———. *Virginia and Tobacco* [pamphlet]. Washington, D.C., 1971.
Trollope, Frances. *Domestic Manners of the Americans*. New York: Alfred A. Knopf, 1949.
Turgenev, Ivan. *First Love and A Fire at Sea*, trans. Isaiah Berlin. New York: Viking Press, 1983.
Twain, Mark. *Huckleberry Finn*. Norwalk, Conn.: Easton Press, 1981.
Tye, Larry. *The Father of Spin: Edward L. Bernays and the Birth of Public Relations*. New York: Crown, 1998.
Van Deusen, Glyndon G. *The Jacksonian Era: 1828–1848*. New York: Harper and Brothers, 1959.
Wagner, Susan. *Cigarette Country: Tobacco in American History and Politics*. New York: Praeger, 1971.
Warfield, Frances. "Lost Cause: A Portrait of Lucy Page Gaston." *Outlook and Independent*, vol. 154, February 12, 1930.
Weinstein, Henry. "Oregon Jury Finds against Philip Morris." *Los Angeles Times*, March 23, 2002.
Wells, H. G. *The Outline of History*, vol. 2. Garden City, N.Y.: Garden City Books, 1961.
Wertenbaker, Thomas Jefferson. *The First Americans: 1607–1690*. New York: Macmillan, 1929.
Wilford, John Noble. *The Mysterious History of Columbus*. New York: Alfred A. Knopf, 1991.
Will, George. *Statecraft as Soulcraft*. New York: Simon and Schuster, 1983.
Williams, Neville. *Elizabeth the First: Queen of England*. New York: Dutton, 1968.
Wilson, A. N. *The Victorians*. New York: W. W. Norton, 2003.
Wilstach, Paul. *Tidewater, Maryland*. New York: Tudor, 1945.
Wineapple, Brenda. *Hawthorne: A Life*. New York: Alfred A. Knopf, 2003.
Winton, John. *Sir Walter Ralegh*. New York: Coward, McCann and Geoghegan, 1975.

INDEX

Ackroyd, Peter, 26
Adams, John, 98, 112
Adams, Samuel, 90
Africa, tobacco introduction in, 34
Airs of Scottis Poesie (James I), 43, 45
All Quiet on the Western Front (Remarque), 162
Allen, Frederick Lewis, 174
Allen, George V., 223
AMA (American Medical Association), 207
Ambrose, Stephen, 85
American Cancer Society, 233, 236
American infrastructure improvement, 154–55
American Tobacco Company, 134, 139, 177, 197, 211, 235
Andagoya, Pascuel de, 20
Anti-alcohol movement, 100–101
Anti-cigarette League of America,
  formation of, 144
  Gaston's role in, 163
  health issues, 152
  legislative support for, 146–47, 149–50
  post World War II campaign, 165–66
Anti-saloon League, 146, 147
Anti-smoking movement, 39
  attitudes and questions, 36–39, 40–41
  end of, 206–7
  James I, role in, 44–45, 68
  roots of, 36
Appointment in Samarra (O'Hara), 174
Argall, Samuel, 63
Aubrey, John, 54

Bain, John Jr., 125
Baker, Newton D., 162
Balboa, Vasco Balboa de, 20
Barnaby Rudge (Dickens), 115
Barrie, James M., 28
Barrymore, Ethel, 158
Bastides, Rodrigo de, 20
Beaumont, Francis, 24
Beeman, Richard R., 78
Belgium, introduction of tobacco in, 32
Benedict XIII (pope), 40
Benzoni, Girolami, 38
Bernays, Edward L., 179, 183, 199, 238
Bernstein, Carl, 223
Black Tuesday, 184
"Blue Law" (1663), 72
Bonsack, James, 134
Boorstin, Daniel, 65, 83
Booth, William, 146
Borthwick, J. D., 108
Boston Tea Party, effects of, 91
Bradford, William, 69
Brice, Fanny, 158
Brill, A. A., 175
Brooks, Jerome E., 26, 47, 114, 123
Brown & Williamson, 230
Bruce, P. A., 61
Bryan, William Jenkins, 165
Bullock, James, 72
Burke, Edmund, 93
Burns, James MacGregor, 83, 165
Byrd, William, 83, 87

Cabral, Pedro Alvarez, 5, 20
Calvin, John, 2
Camel, 180, 195, 199

Cantor, Eddie, 158
Carmen (Bizet), 152
Carnegie, Andrew, 146
Carter, Robert, 83
Carter, Willa, 107
Cartier, Jacques, 13, 20
Casas, Bartoleme de las, 16
The Case Against the Little White Slaver (Ford), 147
Cassavetes, John, 230
Chernow, Ron, 88
Chesterfield, 199
Chicago Anti-cigarette League, 143, 166
Child, Julia, 216
China, introduction of tobacco in, 33, 41
Churchill, Winston, 210
Cigar(s)
  in literature, 107–8
  name meaning, 133
  revival of, 234
  as a social differentiation, 138–39
  social functions of, 11–12, 103
  symbology of, 105
Cigarette(s), 128–35
  first manufacturing firms for, 129
  manufacturing in America, 132
  Natives, relations with, 128–29
  sales, late 19th Century, 134–37
  urbanization, effects on, 137–38
Cigarette Wars (Tate), 175
Clark, William, 12, 73
Clay, Henry, 113
Cohan, George M., 181
The Columbia Cook Book (Hollinsworth), 7
Columbus, Christopher
  death of, 20
  expedition of, 15–18
  tobacco, first encounter with, 16
Common Sense (Paine), 98
Congreve, William, 94
Cooke, Alistair, 92, 110
Coolidge, Calvin, 166, 170, 184
Cooper, James Fenimore, 114
Cosway, Richard, 122
Cotton, 89–90
Counterblaste to Tobacco (James I), 45, 60, 189
Cramer, Richard Ben, 197
Crane, Stephen, 107
Crystallizing Public Opinion (Bernays), 179
Curtis, Tony, 226
Cutlip, Scott M., 184

Dale, Thomas, 57
Darwin, Charles, 169
Davidson, James West, 64
Davis, Bette, 223
Davis, John W., 166
De Bakey, Michael, 193
Dickens, Charles, 115, 223
  in America, 115–16
  on tobacco, 117
Digges, Edward, 85
DiMaggio, Joe, 196
A Discourse Concerning the Plague with Some Preservatives Against It (Byrd), 84
A Discourse Touching a War with Spain and the Protecting of the Netherlands (Raleigh), 51
The Discovery of the Large, Rich, and Beautiful Empire of Guiana with a Relation of the Great and Golden City of Manoa (Raleigh), 50
Doering, Carl R., 192
The Domestic Manners of the Americans (Trollope), 118
Drake, Francis, 21
Draper, Theodore, 93
Duke, James, 134, 150, 171, 178
Dunn, Susan, 165

Ederle, Gertrude, 168
Edison, Thomas, 147
Ehwa, Carl Jr., 61
Eliot, George, 99
Elizabeth I (queen)
  tobacco introduction, role of, 22–23, 41, 42
  Walter Raleigh and, 22–24, 48–49
England
  cigarette factories in, 130
  first anti-tobacco reactions in, 25, 43
  tobacco introduction in, 21–22
Every Man in His Humour (Jonson), 39
(ETS) Environmental Tobacco Smoke, 230

Fairholt, F. W., 31, 42
The Fairie Queene (Spenser), 22, 27
The Father of Spin (Tye), 179
Feodorovich, Michael, 41
Fields, W. C., 158
Finney, Albert, 216
"A Fire at Sea" (Turgenev), 107
Fisher, Sydney George, 80
Fitzgerald, Scott, 169
Fletcher, John, 24
Ford, Henry, 147
Fowler, Orin, 100
Franklin, Benjamin, 97, 124
Freud, Sigmund, 108
Frobisher, Martin, 21

Frum, David, 231
Furnas, J. C., 56, 118, 173

Gabler, Neal, 114
Gama, Vasco de, 20
Gardiner, Edmund, 39
Gaston, Lucy Page, 139, 163, 201, 223
  campaign against tobacco, 141–42, 145–46
  death of, 167
  early life of, 140–41
  post World War II campaign, 165–66
Gately, Ian, 17, 159, 187
Gehrig, Lou, 196
Germany, introduction of tobacco in, 31
Gloag, Robert, 130
Goddard, Robert, 168
Goodman, James, 69
Grant, Ulysses S., 125
The Great Awakening, 76–77
Greeley, Horace, 108
Grijalva, Juan de, 20
The Group (McCarthy), 216

Hamilton, Alexander, 97
Hammond, E. Cuyler, 207
Hancock, John, 110
Harding, Warren, G., 165
Hariot, Thomas, 5
Hawkins, John, 21
Hawthorne, Nathaniel, 108
Heimann, Robert K., 85, 94, 139
Hemingway, Ernest, 169
Henreid, Paul, 223
Henry, Patrick, 70, 76, 77, 98
Herrick, Robert, 27
Hill, George Washington, 177, 178, 180, 182, 196, 211
Hirschfelder, Arlene, 134
The History of the World (Raleigh), 53
Hit Parade, 212
Hollingsworth, Adelaide, 7
Holmes, Oliver Wendell, 155
Hooton, Earnest Albert, 200
Hoover, Herbert, 169, 184
Hopkins, Claude, 170
Horn, Daniel, 207
Hubbard, Elbert, 146
Hume, Noel, 63

An Inquiry into the Effect of Ardent Spirits (Rush), 100
Institutes of the Christian Religion (Calvin), 2
Interstate Commerce Commission, 231
Irving, Washington, 102, 114

Jackson, Andrew, 106, 110, 223
James, Henry, 191
James I (king), 43, 66, 92
Jamestown
  acts of rebellion, 68–69
  formation of, 56–58, 61
  tobacco superstitions in, 64
  tobacco wealth in, 63
Japan, first introduction of tobacco in, 33, 41
Jefferson, Thomas, 73, 85, 89, 95, 98, 122
Johnson, Lyndon, 216
Johnson, Paul, 76
Johnson, Samuel, 39
Jonson, Ben, 24, 39
Jordan, David Starr, 146
Joyful News of Our Newe Found World (Monardes), 29
The Jungle (Sinclair), 156

Kennaway, Ernest L., 223
Kennedy, John F., 216
Keymis, Lawrence, 53
Kieft, Willem, 102
Kipling, Rudyard, 107
Klein, Daniel, 20
Klein, Richard, 94, 131, 160
Koch, Robert, 155
Kool, 195, 212
Kress, D. H., 151
Ku Klux Klan, 169
Kurlansky, Mark, 88

Lacey, Robert, 53
Lamont, Thomas, 184
Langguth, A. J., 91
Lanza, Mario, 195
Larson, Erik, 154
League of Nations, 166
Lee, Robert E., 126
Leon, Juan Ponce de, 20
Levitt, William, 203
Lewis, Carol, 169
Lewis, Jerry, 230
Lewis, John, 77
Lewis, Meriwether, 12, 73
Lickint, Fritz, 194
Lincoln, Abraham, 101, 125, 140, 165
Lindbergh, Charles, 168, 172
Livingston, Philip, 98
L&M, 213
Lombard, Herbert L., 192
London: The Biography (Ackroyd), 26
The Long March (Styron), 161
Longfellow, Henry Wadsworth, 12

Longworth, Alice Roosevelt, 180
Lorillard Tobacco Company, 182
A Lost Lady (Carter), 107
Lucky Strike, 178, 199
Lytle, Mark Hamilton, 64

MacArthur, Douglas, 198
Madison, Dolley, 106
Madison, James, 97
Magellan, Ferdinand, 20
Man Against Crime (commercial), 211
Marlboro, 212
Marlowe, Christopher, 24, 27
Martyr, Peter, 15
Mather, Cotton, 84
Maury, James, 76
Maxim, Hudson, 147
McCarthy, Mary, 216
McDonald, Forest, 70
McKinley, William, 142, 166
McLaren, Wayne, 238
McQueen, Steve, 216
Medici, Catherine de, 30
Mercier, Louis Sebastian, 122
Milton, Giles, 5, 22, 43, 52
Milton, John, 27
Monardes, Nicolas, 29
Monroe, James, 98
Moon, William Least Heat, 214
Morison, Samuel Eliot, 14
Morris, Robert, 98
Muller, Franz, 193
Murphy, Grayson M. P., 157
Murray, Jane, 44

Nation, Carry, 141–43, 145
Neilson, William Allan, 176
Netherlands, first introduction of tobacco in, 32
The New Britannia, 56
Nicholas Nickelby (Dickens), 115
Nicot, Jean, 29, 190
Nietzsche, Friedrich, 169
North American colonies, relations between, 90–91

Obregon, Manrico, 15
"Observations Upon the Influence of the Habitual Use of Tobacco upon Health, Morals, and Property" (Rush), 98
Ochsner, Alton, 193
O'Hara, John, 174
Ojeda, Alonso de, 20
The Old Bachelor (Congreve), 94
The Old Curiosity (Dickens), 115
Old Gold, 195
Oliver Twist (Dickens), 115
O'Neill, Eugene, 169
Otis, James, 98
Ovieda ya Valdes, Gonzalo Fernandez de, 5

Paine, Thomas, 98
Pall Mall, 213
Parkinson, William D., 160
Patterson, James T., 203
Pavarotti, Luciano, 195
Pearl, Raymond, 193
Pemberton, Hannah, 105
Pepys, Samuel, 26
Perrot, Nicholas, 17, 36
Pershing, John J., 157
Persia, introduction of tobacco in, 33
Peter Pan (Barrie), 28
Philip Morris, 195, 199, 228, 236
Picasso, Pablo, 169
The Pickwick Papers (Dickens), 115
Pipe(s)
  social functions of, 11–12, 103
  symbology of, 105
Pocahontas, 60, 66
The Portrait of a Lady (James), 191
Profiles in Courage (Kennedy), 216
Pusey, Joshua, 137
Putman, Israel, 103

Quakers, and smoking, 150
Quincy, Josiah, 90

Raleigh, Walter, 21, 22–24, 27, 30, 39, 47, 91, 106, 222
  death of, 54
  Elizabeth I and, 22–24, 48–49
  England's tobacco introduction, role in, 22, 43
  first El Dorado expedition of, 47–48
  imprisonment of, 52–53
  second El Dorado expedition of, 53–54
Randall, Tony, 226
The Red Badge of Courage (Crane), 107
Redi, Francesco, 189
Remarque, Erich Maria, 162
Revere, Paul, 98
Reynolds Tobacco Company, 170, 235
RICO (Racketeer Influenced and Corrupt Organizations), 237
Robert, Joseph C., 125, 186
Robicsek, Francis, 4, 8
Rogers, Will, 158
Rolfe, John, 60, 71, 92
  tobacco modification, role in, 61

Roosevelt, Franklin, 197
Roosevelt, Theodore, 142
Rosenwald, Julius, 146
Rowlands, Samuel, 27
Rush, Benjamin, 97, 98, 102, 110, 154, 222
  anti-tobacco campaign of, 98–99
Russell, Lillian, 158
Russell, M. A. H., 234
Rydell, Bobby, 216

Sandys, Edwin, 69
Schama, Simon, 32
Schlesinger, Arthur Sr., 88
Shakespeare, William, 24
Shaw, David, 234
Shaw, George Bernard, 148
Shew, Joel, 99
Simmons, James C., 116
Sinclair, Upton, 156
Slader, Matthew, 72
Slavery's presence in North America, reasons for, 87, 90
Slaves, as artisans, 88
Smith, John, 58
Smith, Samuel Francis, 114
Smoke, 9–10
"Smokes for Soldiers Fund," 157
Smoking and Health, 220, 223, 228
Smoking techniques and etiquette, 110, 113
Snuff
  medicinal properties of, 120
  and sneezing, 123
  social function of, 121
  social status, indicative of, 122
Sobel, Robert, 131, 151, 160, 200, 205
"Song of Hiawatha" (Longfellow), 12
Spain
  first cigarette manufacturing firms, 129
  first introduction of tobacco in, 29
Spenser, Edmund, 22
The Spirits of America: A Social History of Alcohol (Burns), 97
Stein, 169
Stella, Benedetto, 40
Stuyvesant, Peter, 69
Styron, William, 161
Sullivan, John L., 147
Sunday, Billy, 164
Swayze, John Cameron, 210

Tabak und Organismus (Lickint), 194
Talman, William, 227
Tate, Cassandra, 166, 175
Taylor, Zachary, 106
Ten O'Clock People, 240–44
Tennant, Richard B., 134, 203
Terry, Luther, 218, 219
Thackeray, William Makepeace, 10, 108
Thorne, William C., 146
Tobacco
  and the age of colonization, 15, 19
  change of role in 17th century, 70–72
  clergy relations with, 73–74
  cost of, 47, 198
  effects of, 4, 9, 65, 194
  in Europe and Asia, 32–33
  first introduction in Europe, 19
  as a "friend of man" perception, 157–58
  Great Depression era in, 184–86
  health issues about, 26, 220–21
  history of, 4
  illnesses caused by, 99
  Jamestown, tobacco production in, 60
  in King James I literary work, 45–46
  in literature, 24–26, 27, 99, 107–8, 161, 191
  as a means of currency and payment, 79, 94, 126
  names for, 30
  natives, relations with, 17, 65–67
  as a patriotic imperative, 95
  research for movies, 207–8
  as a rite of government, 10
  scientific research for, 192–93
  social function of, 79
  superstitions about, 8, 19
  taxes on, 67, 126
  treatments for, 5–6
  use among peasant population, 28–29
  use and application, 7, 16
  use in war, 94
  Vatican reactions from, 40–41
  in World War I, 160
  after World War II, 168–69, 203–5
Tobacco companies
  and entertainment, 229–30
  marketing strategies and sales, 170–73, 179–81, 211, 229, 232
  name changing, 235–36
  rivalry among, 181–84
Tobacco Industry Research Committee, 210
Tobacco Institute, 223
Tobacco plantations
  locations of, 80–81
  slave trafficking in, 85, 87
  slaves in, 80–81
  structure and sizes of, 81–83
  tobacco production, 86
  tobacco sabotages in, 92

"Tobacco War," 95
Tores, Luis de, 16, 17, 36
Trall, R. T., 100
Trask, George, 100
Treaty of Versailles, 164
The Trial of Tobacco (Gardiner), 39
Trollope, Frances, 118, 154, 223
Trotsky, Leon, 167
The True Law of Free Monarchies and Basilikon Doron (James I), 43, 45
Turgenev, Ivan, 107
Turkey, introduction of tobacco in, 41
Twain, Mark, 99
Two-Penny Act, 75
Tye, Larry, 179

Ullmann, Liv, 230
Unique Selling Proposition, 171
Urban VIII (pope), 40

Vanity Fair (Thackeray), 108
Vaquelin, Louis Nicholas, 190
Vatican, tobacco attitude for, 40–41
Verrazano, Giovanni de, 20
Vespucci, Amerigo, 20, 109
Vinton, Bobby, 216
Virginia Slims, 229
The Volstead Act, 164

Wagner, Susan, 189, 208
Wanamaker, John, 146
Wartime
  smoking and, 125–26, 158–59, 199
  tobacco as a means of payment, 73
  tobacco use by soldiers, 94
Washington, George, 94, 157, 222
WCTU (Woman's Christian Temperance Union), 140, 146
Webster, Noah, 114
Whitney, Eli, 89
Will, George F., 112
Willard, Frances, 141
Williams, George, 146
Williams, Roger, 88
Wilson, A. N., 137
Wilson, Woodrow, 157
Winston, 213
Winters, Shelley, 230
Winthrop, John, 88
Women, and smoking, 172–76
Wood, Natalie, 216
Woodward, Bob, 223
Woodward, C. Vann, 185

Xerez, Rodrigo de, 16, 17, 19, 29, 36

Yeardley, George, 64

ERIC BURNS is the host of "Fox News Watch" on the Fox News Channel. He was named by the *Washington Journalism Review* as one of the best writers in the history of broadcast journalism. His other books include *The Spirits of America: A Social History of Alcohol* (Temple), *Broadcast Blues*, *The Joy of Books*, and *Infamous Scribblers*.